Romantic Anti-capitalism and Nature

Romantic Anti-capitalism and Nature examines the deep connections between the romantic rebellion against modernity and ecological concern with modern threats to nature. The chapters deal with expressions of romantic culture from a wide variety of different areas: travel writing, painting, utopian vision, cultural studies, political philosophy, and activist socio-political writing. The authors discuss a highly diverse group of figures – William Bartram, Thomas Cole, William Morris, Walter Benjamin, Raymond Williams, and Naomi Klein – from the late eighteenth to the early twenty-first century. They are rooted individually in English, American, and German cultures, but share a common perspective: the romantic protest against modern bourgeois civilisation and its destruction of the natural environment.

Although a rich ecocritical literature has developed since the 1990s, particularly in the United States and Britain, that addresses many aspects of ecology and its intersection with romanticism, they almost exclusively focus on literature, and define romanticism as a limited literary period of the late eighteenth and early nineteenth centuries. This study is one of the first to suggest a much broader view of the romantic relation to ecological discourse and representation, covering a range of cultural creations and viewing romanticism as a cultural critique, or protest against capitalist-industrialist modernity in the name of past, pre-modern, or pre-capitalist values.

This book will be of great interest to students and scholars of ecology, romanticism, and the history of capitalism.

Robert Sayre is Professor Emeritus of English and American Literature and Civilization at the University of Paris East, Marne-la-Vallée, France.

Michael Löwy is Emeritus Research Director in Sociology at the Centre nationale de la recherche scientifique, Paris, France.

Routledge Explorations in Environmental Studies

Environmental Humanities and the Uncanny
Ecoculture, Literature and Religion
Rod Giblett

Ethical Responses to Nature's Call
Reticent Imperatives
James Magrini

Environmental Education in Indonesia
Creating Responsible Citizens in the Global South?
Lyn Parker and Kelsie Prabawa-Sear

Ecofeminism and the Indian Novel
Sangita Patil

The Role of Non-state Actors in the Green Transition
Building a Sustainable Future
Edited by Jens Hoff, Quentin Gausset and Simon Lex

The Creative Arts in Governance of Urban Renewal and Development
Rory Shand

What Can I do to Heal the World?
Haydn Washington

Towards a Society of Degrowth
Onofrio Romano

Romantic Anti-capitalism and Nature
The Enchanted Garden
Robert Sayre and Michael Löwy

https://www.routledge.com/Routledge-Explorations-in-Environmental-Studies/book-series/REES

Romantic Anti-capitalism and Nature

The Enchanted Garden

Robert Sayre and Michael Löwy

earthscan
from Routledge

First published 2020
by Routledge
2 Park Square, Milton Park, Abingdon, Oxon OX14 4RN

and by Routledge
52 Vanderbilt Avenue, New York, NY 10017

Routledge is an imprint of the Taylor & Francis Group, an informa business

British Library Cataloguing in Publication Data
A catalogue record for this book is available from the British Library

Library of Congress Cataloging-in-Publication Data
A catalog record has been requested for this book

ISBN: 978-0-367-26505-2 (hbk)
ISBN: 978-0-429-29353-5 (ebk)

Typeset in Goudy
by Taylor & Francis Books

For our grandchildren: Kyan, Tania, Esteban, Selma

Contents

Figures

Credits

Images (Chapter 2):

- Figure 2.1: Thomas Cole. The Course of Empire: The Savage State. 1834. Oil on canvas, 39 1/4 x 63 1/4 inches; negative #6045c. ID 1858.1. Collection of The New York Historical Society. Gift of the New York Gallery of the Fine Arts. Digital image created by Oppenheimer Editions.
- Figure 2.2: Thomas Cole. The Course of Empire: The Arcadian or Pastoral State. 1834. Oil on canvas, 39 1/4 x 63 1/4 inches; negative #6046. ID 1858.2. Collection of The New York Historical Society. Gift of the New York Gallery of the Fine Arts. Digital image created by Oppenheimer Editions.
- Figure 2.3: Thomas Cole. The Course of Empire: The Consummation of Empire. 1835–36. Oil on canvas, ca., 51 1/4 x 76 inches; negative #6047c. ID 1858.3. Collection of The New York Historical Society. Gift of the New York Gallery of the Fine Arts. Digital image created by Oppenheimer Editions.
- Figure 2.4: Thomas Cole. The Course of Empire: Destruction. 1836. Oil on canvas, 39 1/4 x 63 1/2 inches; negative #6048c. ID 1858.4. Collection of The New York Historical Society. Gift of the New York Gallery of the Fine Arts. Digital image created by Oppenheimer Editions.
- Figure 2.5: Thomas Cole. The Course of Empire: Desolation. 1836. Oil on canvas, 39 1/4 x 63 1/4 inches. ID 1858.5. Collection of The New York Historical Society. Gift of the New York Gallery of the Fine Arts. Digital image created by Oppenheimer Editions.

Chapters previously published in journals:

- Chapter 1: William Bartram's *Travels* and romantic environmentalism. This chapter is derived in part from an article published by Robert Sayre as "William Bartram and Environmentalism," in *AMSJ: American Studies Journal*, Volume 54, Number 1 (2015), pages 67–87.

- Chapter 5: Raymond Williams: romantic culture and socialist ecology. This chapter is derived in part from an article published by Michael Löwy and Robert Sayre as "Raymond Williams, Romanticism and Nature," in *Capitalism Nature Socialism*, Volume 29, Number 2 (June 2018), pages 75–91. Copyright © 2018 The Center for Political Ecology, reprinted by permission of Taylor & Francis Ltd, http://www.tandfonline.com on behalf of The Center for Political Ecology.

Author biographies

Robert Sayre is a professor emeritus of English and American literature and civilization at the University of Paris East, Marne-la-Vallée, France. He was born in the United States in 1943. He did his graduate studies at Columbia University, and taught as an assistant professor at Harvard before settling in France in 1980. His research areas are diverse and interdisciplinary, involving relations between literature, society, and history. Notable among them are the history and theory of romanticism. He has published over fifty articles and eight books, in some cases collaborating with Michael Löwy on those that concern romanticism. His books include *Solitude in Society: A Sociological Study in French Literature* (Harvard University Press, 1978); *Romanticism Against the Tide of Modernity* (with Michael Löwy; Duke University Press, 2001); and *Modernity and Its Other: The Encounter with North American Indians in the Eighteenth Century* (University of Nebraska Press, 2017).

Michael Löwy is an emeritus Research Director in sociology at the Centre nationale de la recherche scientifique (CNRS). He was born in Brazil to Jewish/Viennese parents in 1938. He has lived in Paris since 1969, and earned two doctorates in France, the first, in 1964, under the direction of Lucien Goldmann. He was awarded the Silver Medal of the CNRS as the best French social scientist in 1994, and has received prizes for his work in Brazil and Mexico. His research interests and publications are wide-ranging, and include notably Jewish messianism, liberation theology in Latin America, surrealism, anarchism, Marx, and Weber, as well as the theory of romanticism that is the basis of the present book. He has published over thirty books, several thematic anthologies and approximately 150 articles, translated into twenty-nine languages. His books include: *Redemption and Utopia: Libertarian Judaism in Central Europe* (Stanford University Press, 1992); *The War of Gods: Religion and Politics in Latin America* (Verso, 1996); *Romanticism against the Tide of Modernity* (with Robert Sayre; Duke University Press, 2001); *Fire Alarm: Reading Walter Benjamin's Theses 'On the Concept of History'* (Verso, 2005); and *Franz Kafka, Subversive Dreamer* (University of Michigan Press, 2016).

Introduction: romanticism, capitalism, and ecology

In this book we do not intend to propose an exhaustive historical study of romanticism and ecology. We have rather selected a series of figures who do not belong to the usual literary canon of romanticism studies, in order to illustrate both the diversity and the coherence of a broad cultural constellation, as well as its continuity well beyond the so-called "Romantic period." The essays contained within it deal with expressions of romantic culture from a wide variety of different areas: literature, travel writing, painting, utopian vision, cultural studies, political philosophy, and activist socio-political writing. We discuss a highly diverse group of figures – William Bartram, Thomas Cole, William Morris, Walter Benjamin, Raymond Williams, and Naomi Klein – from the late eighteenth to the early twenty-first century. They are rooted individually in English, American, and German cultures, but share a common, overarching perspective: the romantic protest against modern bourgeois civilization and its destruction of the natural environment. The purpose of our study is to make visible the deep intellectual, cultural, and emotional connections between the romantic rebellion against modernity and ecological concern with the modern threats to "nature."[1] Moreover, our aim is to show that the essential links between romanticism, anti-capitalism, and ecology can express themselves in very different cultural forms and historical contexts.

Max Weber once observed that Asiatic cultures, with their magical beliefs, live in an "enchanted garden" (*Zaubergarten*),[2] and this concept also can well be applied to the (mainly Western) romantic view of Nature. There exists a rich and interesting literature on romanticism, ecology, and ecocriticism, but most, if not all of it, deals only with literature and only with the so called "Romantic period." Our work is grounded in a radically different concept of romanticism.[3] Far from being consensual, this interpretation goes against the grain of most romanticism scholarship, which is based on the apparently obvious assumption that we are dealing with a literary movement of the late eighteenth and early nineteenth centuries. In our view this assumption is doubly wrong: romanticism is a *worldview* – much more than a literary phenomenon, that is, although of course it has an important literary component – and it did not come to an end either in 1830 or 1848. For us romanticism, as a cultural protest against modern industrial/capitalist civilization, is one of the main forms of modern culture, extending from Rousseau – a particularly crucial foundational figure – up to the present, that is, from the second half of the eighteenth to the beginning of the twenty-first century. Our thesis is based on a (heterodox) Marxist approach to cultural phenomena that attempts to link art, religion, and political ideas to social and historical contexts.

Romanticism as a worldview

What do we understand by "worldview"? We have been inspired by the works of the French cultural sociologist Lucien Goldmann, who expanded on a whole tradition of German thought, notably that of Wilhelm Dilthey. For Dilthey, a worldview (*Weltanschauung*) is an inner form of thinking (*innere Denkform*), that is, a fundamental mindset (*Grundstimmung*). Our approach in treating romanticism as a *Weltanschauung* falls within this tradition, and Goldmann's writings are our starting point although we have reformulated his arguments to a considerable extent. For him a worldview is "a set of aspirations, feelings and ideas that brings together members of a group (in most cases a social class) and opposes them to other groups."[4] Goldmann identified as the main worldviews of the modern era the Enlightenment, romanticism, the tragic, and the dialectical worldviews. Our research on the romantic worldview does not identify it with a single class or group, but with individuals of different social origins, many of whom belong to the social category of "intellectuals," that is, creators of cultural products and representations.

Lucien Goldmann argues that not all bearers of a worldview represent it in an entirely *coherent* way. There are different degrees of consistency and coherence among them. This applies very much to figures who articulate the romantic worldview, including those discussed in this book. Some of them put forward a radical protest against the whole complex of capitalist civilization. They are the ones that most fully incarnate the romantic worldview as we define it. Some others, though, only respond to, react against, or thematize specific aspects of the modern bourgeois world. Some develop a coherent and exclusive romantic perspective, while others vacillate among several perspectives or worldviews, sometimes even mixing them in a single work. Most of those considered as "romantic" by mainstream literary history largely share the romantic viewpoint as defined here. But some belong only partially to it, while others not usually considered as romantic – including those who do not fall within the chronological span of traditionally defined romanticism – can be shown to belong to the romantic style of thinking as we conceptualize it.

In terms of our concept it is also important to point out that non-romantic authors may possess a romantic "moment," aspect, or dimension. A striking example is Karl Marx. Although he was essentially a man of the Enlightenment, Marx's critique of capitalism and his view of history include significant romantic perspectives and arguments, borrowed from writers (Balzac, Dickens), economists (Sismondi), and anthropologists (Morgan, Maurer).[5] A significant number of Marxists in the twentieth century expanded this dimension, and can be characterized as "romantic

Marxists." This includes several of the authors discussed in this volume, beginning with William Morris in the late nineteenth century.

Before defining the romantic worldview in more detail, we should add a note on its relation to societies of the modern period that seem to stand outside capitalism. If romanticism is a protest against capitalist civilization, it would seem paradoxical that it also appears in the so-called "really existing socialist" countries – in the former USSR and other equivalent regimes. For us, however, the decisive point is that the USSR was far from being a socialist society in fact. At best it could be considered as a failed attempt at a transition from capitalism to socialism. It could also be understood, as several dissident Trotskyists such as C.L.R. James proposed, as a kind of "state capitalism." In any case, after a short period of revolutionary experimentation the process of bureaucratization under Stalin's leadership produced a society that had many characteristics in common with Western capitalism: utilitarian rationality, productivism, alienation of work, bureaucratic administration, instrumentalization of human beings, as well as, crucially, the destruction of the environment.

It should be stressed however that there are many fewer important romantic figures among Soviet dissidents than among cultural critics in the West. Aleksandr Solzhenitsyn is a significant example, inspired by a strongly traditionalist and backward-looking form of romanticism, who rejected not only the Soviet totalitarian system but also modern Western European society. While other examples can be found – one example, on the left side of the romantic spectrum, is the East-German author Christa Wolf, to whom we devoted a chapter of *Romanticism Against the Tide of Modernity* – it remains nevertheless that the great majority of romantic writers and artists since the end of the nineteenth century have been in conflict with various manifestations of Western bourgeois industrial society. An interesting case is Aldous Huxley's brilliant novel *Brave New World* (1931), which criticizes modern industrial society of both the East and the West. But his dystopian world, where people do not worship "Our Lord" but "Our Ford," ultimately looks more like Western capitalism than Eastern "socialism."

Romanticism versus capitalism

As we define it more specifically, romanticism is a cultural critique, or rebellion, against capitalist-industrialist modernity *in the name of past, pre-modern or pre-capitalist values*. As a worldview, it is present in a whole range of cultural creations: literature and art, religion and philosophy, political theory, historiography, anthropology, and even political economy. It perceives in the advent of modern bourgeois society a decisive loss of human, social, and spiritual values that existed in a real or imaginary past – the Middle Ages, Homeric Greece, primitive communism, and others.

The romantic protest is always inspired by pre-capitalist values – social, cultural, or religious ones – and by nostalgia for a lost Paradise, a past Golden Age. But this does not mean that it is always reactionary and backward-looking. It can take regressive forms, dreaming of an imaginary return to the past, but also revolutionary ones, which proceed, or attempt to proceed, towards a future utopia via a *detour* through the past. To give an example from one of the authors discussed in this book: William Morris, Pre-Raphaelite poet and artist, admirer of the Middle Ages, came to invest his nostalgia for the past in the revolutionary dream of a communist utopia. These paradoxical, opposing forms of romanticism will be amply illustrated in our study.

The romantic perspective is therefore in direct contradiction with what has been called "the modern regime of historicity," based on belief in the inevitability of "progress" and on rejection of the pre-modern past as "archaic." Referring to our analysis of romanticism as a diversified but broad-based revolt against modernity, the French historian Jérôme Baschet writes:

> It is important to emphasize that the modern regime of historicity did not come to prevail without its reverse [romanticism] also affirming itself … This point is so important that I would propose identifying a romantic regime of historicity … which accompanies [the modern one] as its shadow.[6]

This shadow is precisely its inversion, since the past despised by the modern regime of historicity is claimed by the romantic one, which calls on it in order to criticize the modern present and imagine the future.

As we have already suggested, romanticism does not always challenge the capitalist system as a whole, but reacts often to a certain number of features of modernity that it finds especially odious and unbearable. The following are prominent examples – in a far from exhaustive list – of characteristic, interrelated components of modern civilization frequently lamented or denounced in romantic works:

1 *The disenchantment of the world.* In a famous passage of the *Communist Manifesto* Marx and Engels observed that "the most heavenly ecstasies of religious fervor, of chivalrous enthusiasm, of philistine sentimentalism" of the past had been killed by the bourgeoisie, "drowned … in the icy water of egotistical calculation."[7] Seventy years later, Max Weber noted in a celebrated talk, "Science as a Vocation" (1919):

> The fate of our times is characterized by rationalization and intellectualization and, above all, by the "disenchantment of the world." Precisely the ultimate and most sublime values have retreated from public life either into the transcendent realm of mystic life or into the brotherliness of direct and personal human relations.[8]

Marx and Weber cannot be considered as romantic authors but their descriptions are strikingly relevant. Romanticism may be viewed as to a large extent a reaction on the part of "chivalrous enthusiasm" against the "icy water" of rational calculation and against the *Entzauberung der Welt* – leading to an often desperate attempt to *re-enchant the world.* From this standpoint, the well-known phrase "*die mondbeglänzte Zaubernacht*" (the moonlit enchanted night), penned by the German Romantic poet Ludwig Tieck in 1804, can almost be read as the philosophical and spiritual program of romanticism.

2 *The quantification of the world.* As Max Weber sees it, capitalism was born with the spread of merchants' account books, that is, with the mathematical calculation of receipts and expenses. The ethos of modern industrial capitalism is *Rechenhaftigkeit*, the spirit of rational calculation. Many romantics felt intuitively that all the negative characteristics of modern society – the religion of the god Money (which Carlyle called Mammonism), the decline of all qualitative, social, and religious values as well as of the imagination and the poetic spirit, the tedious uniformity of life, the purely "utilitarian" relations of human beings among themselves and with nature – stem from the same source of corruption: market quantification.
3 *The mechanization of the world.* In the name of the natural, the organic, the living, and the "dynamic," romantic writers often manifested a deep hostility to everything mechanical, artificial, or constructed. They saw the capitalist factory as a hellish place and the workers as damned souls, not because they were exploited but because, as Dickens put it in a gripping image in *Hard Times* (1854), they were enslaved to the machine, to the mechanical movements and uniform rhythm of the steam-engine's piston, which "worked monotonously up and down like the head of an elephant in a state of melancholy madness."[9]
4 *The dissolution of social bonds.* The romantics are painfully aware of the alienation of human relationships, the destruction of the old "organic" and communitarian forms of social life, the isolation of the individual in his egoistic self, which taken together constitute an important dimension of capitalist civilization, centered on urban life. Saint-Preux in Rousseau's *The New Heloise* (1761) is only the first in a long line of romantic protagonists who feel lonely, misunderstood, unable to communicate in a meaningful way with their fellow men, and this is especially the case at the very center of modern social life, in the "urban desert."

Capitalism versus nature

To this list of predominant romantic themes should be added the one that is the central focus of this study: *the destruction of nature*. The waste, devastation and desolation inflicted on the natural environment by industrial civilization is often a deep motive for romantic sorrow and anger. It is intimately related to the previous four objects of romantic protest. Nostalgic for the lost harmony between man and nature, sometimes enshrining nature as the object of a mystical cult, many romantics have observed with melancholy and despair the progress of mechanization and industrialization, the modern conquest of the environment which has led to the disappearance of wilderness and the defacing of beautiful landscapes. The poisoning of social life by money and the poisoning of the air by industrial smoke are understood by some romantics as parallel phenomena, stemming from the same perverse root – the ruthless domination of utilitarianism and commercialism, the dissolving power of quantitative calculation. In the disenchanted capitalist world, nature ceases to be a magical, spiritual realm, a divine creation, or the sacred splendor of beauty. Forests, rivers, and landscapes are reduced to raw materials, only to be exploited until exhaustion.

Romanticism and nature: The origins

The origins of what we call the romantic "enchanted garden" can be found among the early romantics, that is to say the writers and philosophers usually identified as romantic. Although for us the romantic worldview is not limited to the so-called Romantic period but is alive in modern culture up to the present, the early romantics were undoubtedly the ones who laid the first stepping-stones of the unfinished romantic narrative. Romanticism of course does not have a single birth date. But if we wanted to pick one moment as a symbolic starting point it might be 1755, the year in which Jean-Jacques Rousseau published his *Discourse on the Origin and Basis of Inequality among Men*. This astonishing document constitutes perhaps the first romantic manifesto, with its ferocious critique of modern civilization and celebration of the "noble savage." The association between these two *topoi* is to be found among many later romantic writers and artists, from the eighteenth century up to our time, as will be evident throughout this book.

While Voltaire, the great proponent of Enlightenment and progress, pictures indigenous peoples as anthropophagic barbarians in his philosophical satire *Candide* (1759), the romantic Rousseau sees them as "the true youth of the world." For him, all the subsequent steps of progress, which were supposed to lead towards the perfection of the individual, "were in effect steps towards the degradation (*décrépitude*) of the species." The savage "breathes only rest and freedom," while civilized man "works until he dies" and is "proud of his serfdom."[10] In fact, Rousseau emphasizes, the barbarian "refuses to bow his head to the yoke that the civilized person bears without grumbling," and prefers the most dangerous freedom to the most peaceful submission. In a passage that seems almost to foresee anticolonial struggles, Rousseau argues that the love of freedom is so strong among "savages" that they "are willing to face hunger, fire, iron and death to keep their independence."[11] Although the philosopher's "state of nature" may be a fiction, his portrayal of the life of primitive peoples is almost surely based on travelers reports. Rousseau in any case often explicitly refers in his essay to specific groups: Hottentots, Caribbeans, and "savages of the Americas."[12]

In the *Discourse* Rousseau also denounces modern destructive behavior towards the natural world. He admires "the immense forests that the felling axe never mutilated," and regrets that civilization has made of the human being a "tyrant over himself and over nature."[13] Worried that the expansion of agriculture might lead to "the destruction of the soil," of its fertility, that is, he quotes a passage from Buffon's *Natural History* (1752) which seems almost prophetic:

> Since men consume enormous quantities of wood and plants for fire and other uses, it follows that the layer of vegetal earth in an inhabited country must always diminish and finally become like the terrain of Arabia Petrae, and like that of many other provinces of the East – that is, in fact, the regions of most ancient habitation – where only Salt and Sand are found.[14]

Another essential romantic aspect of Rousseau's writings is a passionate, quasi-mystical relation to nature. In his *Reveries of a Solitary Walker* (1778) he describes his "rapture" and "extasy" when confronted with the marvelous spectacle of nature. The more sensitive the soul of the observer is, the more "a sweet and deep reverie takes over all his senses, and he loses himself in a delicious intoxication (*ivresse*) in the immensity of this system, with which he feels identified." In the midst of trees and other green things the writer exclaims: "I feel I am in the earthly paradise."[15] Unlike some other authors we will discuss, Rousseau does not generally connect these two romantic moments – love of nature and admiration for the "savage" way of life. In Note IX of the *Discourse on the Origin and Basis of Inequality among Men*, however, one ironic passage does seem to relate the two, in contrast to "civilized" behavior: "The savage, after having his dinner, lives in peace with Nature and in friendship with his fellow creatures."[16]

The example of Rousseau illustrates the fact that the romantic attitude to nature cannot be reduced to a purely aesthetic one, as popular stereotype would sometimes have it. Discussing *Naturphilosophie* in the writings of Novalis, Schelling and J.W. Ritter, the Greek scholar Stephanos Rozanis points out that this romantic philosophy – which is also a sort of theology – has as its supreme value the spiritualization of nature. The natural cosmos is seen as divine, and divinity by the same token is conceived of as the soul of the cosmos, the "*Weltseele*," as Schelling terms it.[17]

Early romantics in Europe at the end of the eighteenth and beginning of the nineteenth centuries indeed often saw nature as a sacred and magical universe, the expression of a divine spirit. As François-René de Chateaubriand wrote in his *Génie du christianisme* (1802), "The gift of prophecy and wisdom, of mystery and religion, seems to dwell eternally in the sacred deepness of the forests." Indeed, Chateaubriand announces as his overall ambition in writing *Génie du christianisme*, nothing less than "to oppose a *religious natural history* to the modern scientific books."[18] For many of the early romantics, especially in Germany, nature constituted a kind of mineral and vegetal metaphysics, a cabbalistic secret language to be deciphered, and through their philosophy of nature they aimed both at the naturalization of the spirit and the spiritualization of nature.[19]

They also often perceived nature to be a kind of arcadian garden of Eden. The German romantic *Naturphilosoph* Gotthilf Heinrich Schubert regrets the "first times, when our species lived in deep harmony with the whole of Nature," an era of "spiritual peace and paradisical joy" which the ancient nations celebrated as the Golden Age.[20] This leads them to a critical view of the modern destructive relation to the natural world, as for instance in Novalis's unfinished novel, *The Disciples at Sais* (1802). Depicting the attitude of a promethean spirit bent on conquering and submitting nature by all means, Novalis ascribes to him the following aims: "Let our generation wage a slow and well-thought-out war of destruction (*Zerstörungskrieg)* against this Nature. We must dominate her by crawling poisons." For Novalis, this conquering attitude results in nature being seen as an "enraged monster" that should be "forever paralyzed" by human action, thereby putting "an end to its devastations."[21] With these and similar comments by German writers at the beginning of the nineteenth century, the romantic attitude to nature ceases to be only an aesthetic and religious feeling – or, in some authors, a highly speculative philosophical exercise – to become a substantial *Zivilisationskritik.*

Another German writer of a very different kind was deeply influenced by the spirit and milieu of early German romanticism: the natural philosopher and scientist Alexander von Humboldt. Before undertaking his extensive travels in South America and elsewhere which led to a series of ground-breaking works enunciating and illustrating a new vision of the natural world, Humboldt spent several years in the mid-1790s in Jena and Weimar, where he frequented the romantic circle of artists and intellectuals around Goethe and Schiller. Humboldt was particularly marked at this time by the organicist *Naturphilosophie* of Friedrich Schelling, then a professor at the University of Jena and a member of Goethe's circle. The romantic views of nature and man's relation to it articulated by Schelling and others of this milieu were important formative impulses in the later development of Humboldt's own conceptions. While his intellectual and cultural roots were thus in German romanticism, Humboldt in turn had a decisive impact on the views of nature held by the English romantics Coleridge and Wordsworth, most especially the former, and later on those of the Americans Thoreau and Emerson. In this way he provided one of the crucial bridges between German and Anglo-American romanticism in terms of perspectives on the natural world.[22]

Although Humboldt's view of nature did not carry over the religious dimension of German romanticism, it in many other ways reflected the latter perspective. Unlike the romantics of the Jena circle, Humboldt was first and foremost a natural scientist devoted to the study of natural phenomena through observation and measurement. But his approach to nature was exceptional in that it joined indissolubly the scientific and the "humanistic." At the precise historical moment when "science" was separating itself from philosophy and becoming increasingly specialized, Humboldt articulated a broadly holistic vision of nature that included the sensual, emotive, imaginative, and artistic responses of the human subject in relation to it. In his multivolume account of his South American travels Humboldt exclaims in one passage: "Nature everywhere speaks to man in a voice … familiar to his soul."[23] The artist is particularly attuned to this correspondence, and to the beauties of nature, and in another important work, *Views of Nature* (*Ansichten der Natur*, 1808), Humboldt writes concerning the painter of natural scenes, that "under his hand the great, magical image of Nature (if I may venture to use the expression) reveals itself … in a few simple strokes."[24] Indeed, as Andrea Wulf emphasizes in her biography, throughout the writings of Humboldt one finds, side by side with scientific observation and analysis, passionate expressions of a sense of wonder and sensitivity to the "magic" in the natural world.

Humboldt is now increasingly recognized as a genial precursor of ecological science in his global and interrelational approach to this world, seen as a vast web of connections covering the entire planet, and more specifically in his theorization and study of "climate zones" and climate change.[25] But equally important, from our point of view, is his "enchanted" perception of Nature, accompanied by a wide-ranging and penetrating critique of the civilization of modernity and its deleterious effects. Although he admired the political ideals of the early United States, he was harshly critical, when he visited the country, of the evils he observed there: slavery and the expropriation of the lands of Amerindians, both stemming from commercial imperatives (Humboldt had studied finance before turning to science, but detested it), and the pervasive mercantile mentality that made of that nation, as he wrote to a German friend, "a Cartesian vortex, carrying away and levelling everything to dull monotony."[26]

In addition to this broad criticism of the evils of modernity as discovered in the United States, in many places in his writings Humboldt also points to the disastrous effects on the natural environment of this civilization, often motivated by greed. As he observed throughout his travels – in South America, Europe, and Russia – mining, modern forms of agriculture that include monocultures and intensive irrigation, as well as early industry, often lead to severe deforestation, impoverishment of the land, and pollution.[27] In his travels he also observed the traces of earlier, pre-modern cultures, strikingly different in their ethos and in their relation to the environment. His comments on these cultures manifest a strong interest and sympathy – another aspect of his sensibility that he shares with many other romantics.[28]

The last figure that we will mention here from the early period at the root of the romantic revolt is the English poet John Clare. In some ways Clare seems to stand apart from the lines of development we have been discussing, and in stark contrast to Humboldt specifically. Humboldt was a wealthy, highly cultivated aristocrat, while Clare was a self-educated rural worker. Humboldt was a world traveler, whereas Clare remained for almost his entire life in the narrowly circumscribed region of his birth. In view of this relative isolation, it is not surprising that Clare was not one of the English romantics for whom Humboldt provided a bridge to German romanticism. Nor does he apparently share much in common with the early French romantics like Rousseau and Chateaubriand. Nonetheless, Clare expresses, in his own idiosyncratic terms, and very powerfully, the nature-inflected romantic vision such as we conceive it.

Clare's rebellion against the civilization of early capitalism takes the form, specifically, of an intense, personal reaction to the experience of the enclosure of rural lands that took place throughout England at the beginning of the nineteenth century – the parceling up of lands previously held as commons into privately owned lots with physical boundaries, and the "improvement" of rural property by clearing wilder, previously common tracts of land and introducing modern agricultural practices. In the area around Clare's home town, Helpston in Northamptonshire, the process took place over a little more than a decade – 1809 to 1820 – a period which corresponded to his coming of age (he was sixteen in 1809). He experienced the transformation of his native countryside, and of its inhabitants' way of life, as an unmitigated disaster on several levels, one of which was socio-economic. Coming from a poor, landless family, Clare struggled simply to survive, and ironically was forced by necessity to engage himself as a worker on some of the very projects he despised.[29]

Throughout the period of the Helpston enclosures and after they had been completed, Clare wrote a series of poems on the theme that have been called "enclosure elegies."[30] In them he lovingly evokes the natural surroundings and lifeways of his area before enclosure, describes the effects wrought by the latter and expresses a mixture of grief, melancholy, and anger at the result. Taken together, they forcefully articulate a romantic anti-capitalist vision that is focused on nature and the place of human beings within it.

In these poems[31] the poet reveals an overwhelming nostalgia for the meadows, moors, and bogs that he roamed as a child. These partially wild lands are characterized importantly by the freedom they allow both to humans and to other living things:

> Unbounded freedom ruled the wandering scene / Nor fence of ownership crept in between … Now this sweet vision of my boyish hours … / Is faded all …[32]

The lands are portrayed as beautiful as well, although the esthetic dimension of nature is only one among several in the poems. Of primary importance is the collective, social aspect of human community in nature, and Clare's nostalgia goes far beyond the personal joys of childhood play. As commons, the land provided support for the poor and propertyless, and was also the site of forms of premodern community, in particular that of gypsies. Clare enjoyed frequenting a gypsy encampment at one of his favorite places in the local commons; several poems mention it, and lament its eviction by enclosure. More generally, Clare was strongly attracted to folk cultures, and collected folk music, stories, and the like.

In the "enclosure elegies" Clare protests vehemently against the demolition, by enclosure and modernization, of both wild natural sites and the premodern cultures that found a home on them, and clearly identifies the agency behind the change. In the earliest of these poems, "Helpston," he writes: "Accursed wealth oer bounding human laws / Of every evil thou remains the cause," and more specifically links it to both the suffering of poor laborers and damage to nature.[33] Later poems point to "self-interest" and the seeking of "profit" or "gain" as the motives propelling the damaging changes; those who are so motivated have "little minds," and justify their depredations with the strident doctrine of economic freedom (contrasted by the poet with the true freedom afforded by nature).[34]

Although Clare's politics were nominally conservative, his poetic indictment of the incursions of modernity in a romantic perspective is thus a radical one. The force and pertinence of his critique has indeed been recognized and honored by later critics who share an affinity with him: E.P. Thompson, who wrote a tribute on the bicentenary of his birth, Raymond Williams, who published, with his daughter, an anthology of Clare's poetry, and most recently George Monbiot, the Guardian's ecological columnist, who emphasizes in an article celebrating Clare the relevance for today's environmental crisis of his poetry of protest.[35]

Is it legitimate to call "ecological" the romantic approaches to nature that we have looked at the origins of here, and that we will explore further in the body of our work? Not perhaps in the limited scientific sense of the word "ecology," as defined by Ernst Haeckel, its inventor. Nor in the sense of a modern social movement fighting against the negative environmental consequences of capitalist modernity, although in some cases – several of which are discussed in our work – romantics have become engaged in these. But as we will attempt to illustrate in this book, one finds in the romantic current of cultural critique a sui generis form of consciousness that is ecological in the most meaningful sense, one that has played an essential part in the historical development of ecology and that continues today to be a potent force in ecological protest and activism. We might perhaps define the common attitude of the figures discussed in this book as "romantic ecocriticism" – not in the more limited meaning most often attributed to the latter term, but more broadly as a radical cultural and moral revolt against the damage resulting from the interaction of modern human societies with nature, in the name of qualitative values lost in modernity.

Romanticism and environmental humanities

We hope that our study of the ecocritical current in romantic anti-capitalism can contribute to the ongoing trend that has been broadly labeled "environmental humanities." The first defining characteristic of this movement is of course the integral connection between environmental issues and concerns addressed by the humanities, that is, questions involving cultural phenomena. More specifically, this approach understands ecological crises as crises of culture. Our exploration of romantic valuing of the natural world, and protest against the destructive effects on that world of modern industrial capitalism, is a major instance, in the context of Western civilization, of the intimate interaction between environment and culture.

Another important trait of the environmental humanities overall is its interdisciplinarity, with the various sub-areas of the trend bringing different disciplines together. Our study, which falls within the domain of environmental cultural studies, brings together a wide diversity of cultural phenomena as expressive of ecocritical romantic anti-capitalism. They include travel literature, landscape painting, utopian writing, social philosophy, literary and cultural studies analysis, and the socio-political essay. In discussing these diverse cultural expressions we continually cross disciplinary boundaries, and our concept of romanticism, which has historical, sociological, economic, and cultural dimensions, is also explicitly interdisciplinary.

It has been pointed out that what the environmental humanities do, in relation to the traditional humanities, is among other things to effect an *enlargement* of perspective. Questions which the traditional humanities often treated within restricted philosophical or literary bounds, are opened onto a larger, contextual frame. This is indeed what our conceptualization of romanticism attempts to do, since it situates expressions of the romantic worldview as critical responses to socio-economic developments in modernity, which crucially include increasing devastation of the environment.

One strong theme that is often articulated in work done within the environmental humanities is affirmation of the unity of nature as an organic/inorganic totality, one that takes the form of a complex network of links. This conception is generally shared by the romantics that we discuss, and more broadly it can be said that the environmental humanities' understanding of natural oneness and interrelation has close affinities with prevalent romantic views of nature.

Another pronounced tendency within environmental humanities has been to explore, and recognize the value of, other conceptions of nature than the dominant, modern Western paradigm. In addition to exploring Eastern thought concerning the environment, there has been considerable interest in "indigenous" peoples' views and practices. In our study we give a prominent place to these, showing how they relate to other romantic ecological understandings, and pointing to their particular importance in the contemporary context.

In addition to these general connections, we should briefly indicate the ways in which each of the individual figures that we discuss brings into play ideas and perceptions that are relevant to discussions taking place within the environmental humanities:

Chapter One: William Bartram affirms the non-hierarchical unity of life forms, emphasizes the "dignity" of animal nature and is sharply critical of cruelty to animals. His botanical sketches illustrate the interconnectedness of the organic and inorganic forms depicted.

Chapter Two: Thomas Cole warns apocalyptically, both in writing and painting, of the irreparable human and natural damage that will occur, unless the process is stopped, through modern incursions in, and destruction of natural sites, impelled by a profit-oriented utilitarianism.

Chapter Three: William Morris denounces modern civilization's approach to nature as a conqueror standing outside it, rather than as an integral part, living harmoniously within it. His *News From Nowhere* imagines a future "ecotopia" in which the harmony of earlier societies is restored at a higher level and the split between country and city transcended.

Chapter Four: Walter Benjamin mounts a radical critique of the "plundering" of nature in capitalist modernity, going so far as to define its relation to nature as "murderous." He shows sympathy with non-destructive attitudes towards nature exhibited by "primitive" societies, and warns of impending ecological, as well as human disaster unless the "emergency brake" is pulled.

Chapter Five: Raymond Williams, like William Morris, calls for overcoming the opposition between country and city, criticizes the ideology of "modernization" as continual, beneficial "progress," and sees the need to reconceptualize the notion of "production," broadening it to include by-products – environmental damage in particular – as well as products. As a socialist thinker, he points to the shared exploitative attitude towards nature held by both capitalist and so-called communist societies, and makes a plea for a radically different "green socialism."

Chapter Six: Naomi Klein, similarly to Williams, feels that "ecological deficits" should be measured along with economic growth, and strongly opposes the "extractivist mind-set" that predominates in the contemporary world of globalized capitalism. She sees a clear connection between the ongoing ecological crisis and the endangerment of peoples "othered" by dominant elites, notably the indigenous inhabitants of white settler societies. She has been greatly impressed by the participation of native peoples in contemporary struggles to limit environmental damage, and admires their cosmological traditions that see all living creatures as "relations," as well as taking a stance of "stewardship" in relation to the natural world.

Recent work on romanticism and ecology

We are by no means the first to have explored the links between "romanticism" and "ecology" but, as will be seen in the short survey that follows, most studies up to now have dealt almost exclusively with the literary aspect of this connection. Our study – one of the first to suggest a much broader view of the romantic relation to ecological discourse and representation – clearly stands apart from them in this way.

Contemporary recognition and analysis of the connection between the two terms is closely related to, even nearly identical with the development of "ecocriticism" in the more restricted meaning to which we have alluded. Sometimes also called "green studies" or "green reading," this critical approach to literary texts first emerged in Britain and the United States, though it has more recently spread to other countries. Defined in an early anthology simply as "the study of the relationship between literature and the physical environment,"[36] it has evolved and diversified considerably since its inception. Although several isolated works appeared in the 1970s, the real take-off of the approach took place in the 1990s, with the publication of a number of seminal monographs as well as *The Ecocriticism Reader* just cited, and the creation of the Association for the Study of Literature and Environment in 1993.[37] In the twenty-first century ecocritical literary analysis has burgeoned exponentially, with some half dozen works published in the year 2000 alone, and dozens more having appeared since then. These later works, often referred to as the "second wave" of ecocriticism, have often problematized key concepts and brought other forms of critical discourse (post-colonial, post-modern, feminist, etc.) to bear on ecocritical discussions.[38]

But a common denominator of ecocritical studies throughout has been a primary focus on authors traditionally identified as romantic. In most cases the definition of romanticism that is at least implied is the usual one of a literary period, and since the ecocritics are generally Anglo-American, the overwhelming emphasis has been on English and American literary writers of the late eighteenth and early nineteenth centuries. Although there has been some debate about which of the early romantic authors most fully express an ecological viewpoint, a large consensus has concluded that important roots of modern ecology are to be found in the literature of the Romantic period. This filiation is suggested, for example, in the subtitle of *The Green Studies Reader* – "From Romanticism to Ecocriticism" – the first section of which includes excerpts from, or discussion of, Blake, Wordsworth, Coleridge, Thoreau, and Clare. Included also in this reader are passages from John Ruskin and William Morris, and indeed these and other late nineteenth century figures are often part of ecocritical genealogies that begin in the Romantic period. The precise relation of the authors with romanticism is not however usually specified. In some cases the filiation is carried into the twentieth century, and authors of that century occasionally are identified with romanticism.[39] For the most part, though, romanticism is seen more as an origin than as a continual presence in ecological thought and representation.

In the literature of ecocriticism one finds a sustained discussion of the history of ecological consciousness, thought and representation, and its relation to romanticism. In this discussion some debates and differences of perspective have emerged. Most of those analyzing the development of modern ecology in some way distinguish between two strands of its history, using various terms to describe them: on the one hand the "spiritual," "humanistic," and "subjective" approach, on the other the "scientific," "rational," and "objective." One way of construing the historical relationship between the two strands is to see the second as originating only in the latter half of the nineteenth century, especially in the work of the German zoologist Ernst Haeckel, who coined the term in 1866.[40] From this point of view, the first strand precedes the second chronologically, and is sometimes seen to be "proto-ecological" rather than to fully meet criteria for the ecological perspective.[41] While the "first wave" ecocritics often stressed the distinction between the two strands of ecological discourse, and emphasized the crucial contribution of the early literary/philosophical – and romantic – writers in their difference from the scientific approach, "second wave" critics have increasingly questioned the opposition, pointing out that literary writers of the Romantic period, at a time before clear intellectual and institutional separations had developed between literature, philosophy and science, often were cognizant of, and influenced by, the writings of contemporary naturalists and "natural philosophers." Far from being out of touch with inquiries into the natural world, romantic literary authors' work was often irrigated by them.[42]

Histories of ecology as set out in the ecocritical literature are in general agreement, though, that its beginnings are to be found at the turn of the eighteenth to the nineteenth century, in association with early romanticism, and that it evolves afterwards through a number of new phases or turning points: the enunciation of a specifically "scientific" approach in the 1860s; the conservationist movements of the late nineteenth and early twentieth centuries; the new awareness of ecological threats in the 1960s and 1970s (Rachel Carson's immensely influential *Silent Spring*, which warned of the danger of pesticides, appeared in 1962); and finally the widespread awakening to the immanence, severity, and global nature of ecological crisis in the 1990s and beyond.

Although studied and chronicled most intensively in the Anglo-American sphere, since ecocriticism has largely come out of England and the US,[43] manifestations of these developments occurred in other countries as well, notably in Europe. In his recent history of ecology (*Die Ära der Ökologie*, 2011; English translation: *The Age of Ecology*, 2014), when discussing origins the German scholar Joachim Radkau points to the contribution of the German *Sturm und Drang* movement (alongside Rousseau), as well as to a significant cultural trend in early nineteenth century Germany that he calls "forest romanticism."[44] He also alludes to the later ecological awareness of the founding fathers of Marxism (although it was not one of their primary themes), citing Engels's comment in *The Dialectics of Nature*: "Let us not, however, flatter ourselves overmuch on account of our human victories over nature. For each such victory nature takes its revenge on us."[45] As for the French history of ecological thought and imagining, Serge Audier's recent broad panorama traces a French tradition that includes the novelist George Sand, the utopian thinkers Fourier and Proudhon, as well as the romantic historian Jules Michelet.[46] Audier and others have also pointed to several significant Russian contributions to the growth of ecological consciousness – notably in the work of the anarchist Piotr Kropotkin and the back-to-the-land movement around Leo Tolstoy.

The range of socio-political positions of those associated with ecology is very wide. The editors of the recent anthology *Ecology and the Literature of the British Left* acknowledge the strong imprint on ecology of positions from the right due to the "essentially conservative and often pessimistic impulse behind much environmental concern," as well as the tendency of much socialist thought to espouse the principle of unlimited growth;[47] but the aim of their volume is to explore the rich interaction and interpenetration of the "red" and the "green," demonstrating through the various contributions a lineage running from early British romantics, through Ruskin and Morris, to later expressions of "eco-anarchism" and "eco-socialism."

In fact, we can notice a clear parallel between romanticism and ecology in this regard. In our study of romantic anti-capitalism – *Romanticism Against the Tide of Modernity* – we argued that that worldview is politically "hermaphroditic," crossing all political boundaries, and that this characteristic accounts, in some cases, for radical shifts in political orientation over a particular author's career, while the fundamental romantic vision remains unchanged.[48] The same might be said, *mutatis mutandis*, of ecological concern and commitment. This heterogeneity of socio-political stances in both romanticism and ecology is naturally reflected, then, in the present study of romantic "ecologists," though in it the left-leaning "red" thread of political orientation is predominant.

As the above brief overview has attempted to suggest, a voluminous and rich literature has developed since the 1990s that addresses many aspects of ecology and its intersection with romanticism. In this literature, a number of contending arguments have been put forward. The most fundamental perhaps involve the definition of two key concepts in the field: "nature" and "ecology." Although discussions around these terms are certainly not without interest, we will not choose among the possible options but rather adopt a position that includes all within the frame of our analysis. On the question of what "nature" is, some have proposed distinguishing "nature" (the brute physical reality) from "Nature" (the human cultural construct); others have proposed abandoning the concept altogether and replacing it with the notion of "environment;"[49] others still have contested the idea of a non-human natural world, and redefined nature as the "more-than-human" world. These distinctions and questionings, however interesting in other contexts, are sometimes at odds with each other and at any rate do not seem directly relevant to our exploration of romantic expressions of ecological awareness and protest. Thus we will, as stated earlier, adopt a broadly inclusive, commonly recognized understanding of nature and the natural world.

Similarly, the ecocritical literature has proposed an array of defining characteristics, some of them contradictory, of the ecological perspective or "outlook." They would include: a sense of the interrelatedness of natural phenomena, in a "mesh" or "ecosystem;" a treatment of nature as "for-itself" (versus for humans), also sometimes called "ecocentrism;" a conception of humans as integral to nature, or, on the contrary, a recognition of the non-human as radically "other;"[50] a sense of place, of particular natural locales, or, on the contrary, a more universal gaze; a secularized, physical apprehension of nature, or, on the contrary, its sacralization. The list could be extended, but the point we wish to make is that any or all of these traits can be found in romantic ecological sensibilities.

In this book we hope to shed new light on the subject of romanticism and ecology by reframing it in terms of our interpretation of romanticism as an anti-capitalist *Weltanschauung* of the entire modern period. We broaden the scope of romanticism here in two principal ways. First, taking it beyond the traditional period definition, we identify and treat as romantics all of the figures dealt with, beginning with William Bartram (second half of the eighteenth century) and ending with Naomi Klein (contemporary, twenty-first century). Second, we illustrate the current of romantic ecology with examples that break away from the usual literary focus. William Bartram was a naturalist and botanical artist, Thomas Cole a landscape painter, William Morris an artisan and artist working in multiple media, Raymond Williams a literary and cultural critic, Walter Benjamin a multi-faceted social philosopher, Naomi Klein an essayist and activist. Although a number of them – Cole, Morris, and Williams – did produce some literary works, they are not the principal creative activity of these figures.

The most important innovation, perhaps, of our approach in relation to the ecocritical literature, is the linkage of the ecological romantics that we discuss with anti-capitalism. Ecologists and scientists recently have come to the widespread agreement that we have entered a new geological epoch, the Anthropocene, in which human activity has changed essential aspects of the environment, ultimately with dire consequences. Many "eco-Marxists" – Ian Angus and John Bellamy Foster, in particular – accept this concept, but emphasize that it is a specific organization of human activity that is responsible for these disastrous changes: the modern capitalist mode of production with its civilizational correlates.[51] This point is forcefully and explicitly made by Naomi Klein, the subject of our last chapter, but we attempt to show how this awareness, in one form or another (including in periods before the word "capitalism" was in usage), is to be found in all of the figures we explore. And, crucially, we wish to highlight the *critical* dimension of that awareness. In this sense, then, we suggest that the term "romantic ecocriticism" can be aptly used to describe the form of romantic anti-capitalism we have investigated here.

Notes

1 Although we recognize that its meaning has recently been much debated, we will not enter into those discussions here and will use the term "nature" in its widely accepted sense of the biophysical universe, and in particular the various forms of life on our planet.
2 Weber, *Gesammelte Aufsätze zur Religionssoziologie II*, 371.
3 For a systematic discussion and illustration of our conception of romanticism, inspired by the writings of Georg Lukács, Ernst Fischer, and others, but from a different perspective, see our book-length study *Romanticism Against the Tide of Modernity*; see also our essay, "Romanticism and Capitalism," in *A Companion to European Romanticism*.
4 Goldmann, *Le dieu Caché*, 26. (Translation ours. Unless otherwise indicated, translations in this book from works in foreign languages are ours).
5 For an extended discussion on Marx and romanticism, see Löwy and Sayre, *Romanticism Against the Tide of Modernity*, Chap. 3.
6 Baschet, *Défaire la tyrannie du present*, 66. Baschet draws on the concept of "regimes of historicity" as it has been elaborated notably by François Hertog.
7 Marx and Engels, *Manifesto of the Communist Party* (1848), 487.
8 Weber, "Science as a Vocation," 302.
9 Dickens, *Hard Times*, 22.

10 Rousseau, *Discours sur l'origine et les fondements de l'inégalité parmi les hommes* (1755), 118, 146.
11 Rousseau, *Discours sur l'origine et les fondements de l'inégalité*, 132–133.
12 Rousseau, *Discours sur l'origine et les fondements de l'inégalité*, 78, 147.
13 Rousseau, *Discours sur l'origine et les fondements de l'inégalité*, 70, 80.
14 Rousseau, *Discours sur l'origine et les fondements de l'inégalité*, Note IV, 154–155.
15 Rousseau, *Rêveries du promeneur solitaire* (1778), 98, 124, 151.
16 Rousseau, *Discours sur l'origine et les fondements de l'inégalité*, 163.
17 Rozanis, *I Romantiki Exegersi* (The Romantic Insurrection), 34–35, 41.
18 Chateaubriand, *Le génie du christianisme*, Vol. 1, 316, 157.
19 Fischer, *Ursprung und Wesen der Romantik*, 234, 238.
20 Schubert, "Ansichten von der Nachtseite der Naturwissenschaft" (1808), 76–77.
21 Novalis, *Werke*, 286–87.
22 On the relationship of Humboldt with the Jena circle on the one hand, and with Anglo-American writers on the other, see Andrea Wulf's remarkable intellectual biography: *The Invention of Nature*, 5, 25–33, 36, 72, 128–29, 167–71, 250, 256–57.
23 Humboldt, *Personal Narrative of Travels to the Equinoctial Regions of the New Continent during the years 1799–1804* (1814–29), Vol. 3, 160. Cited in Wulf, *The Invention of Nature*, 54.
24 Humboldt, *Views of Nature*, 168, and cited in the introduction, 8–9.
25 See the introduction to Humboldt and Bonpland, *Essay on the Geography of Plants*, 2–5.
26 Cited in Wulf, *The Invention of Nature*, 3. See also 19–20, 106, 108, 181, 276.
27 See Wulf, *The Invention of Nature*, 56, 58, 103–05, 213, 288.
28 Although Humboldt's comments on indigenous cultures occur in many of his writings, a series of commentaries, along with pictorial representations, are to be found in *Vues des Cordillères et monumens des peuples indigènes de l'Amérique* (Views of the Cordilleras and Monuments of the Indigenous Peoples of the Americas; 1810).
29 On Clare's relation and response to the enclosures in Helpston, see Goodridge, *John Clare and Community*, Chap. 5.
30 See Goodridge, *John Clare and Community*,105.
31 Some of the best examples are "The Fallen Elm," "Remembrances," "The Lament of Swordy Well," and "The Mores" (Moors).
32 Clare, *Selected Poems*, "The Mores", 169–70.
33 Cited in Goodridge, *John Clare and Community*, 106.
34 See Clare, *Selected Poems*, 168, 170, 172, 197.
35 E.P. Thompson's brief celebration of Clare appeared in a special bicentenary issue of *The John Clare Society Journal*, No. 12, July 1993, 31; the anthology edited by Merryn and Raymond Williams is: *John Clare: Selected Poetry and Prose*; the Monbiot column on Clare is : "John Clare, the poet of the environmental crisis – 200 years ago," *The Guardian*, July 9, 2012.
36 *The Ecocriticism Reader: Landmarks in Literary Ecology*, xviii.
37 On this early period of the trend, see Pinkney, "Romantic Ecology," 411–12.
38 For a good overall discussion of later developments in the field, see Rigby, "Romanticism and Ecocriticism."
39 See, for example, Tovey, *The Transatlantic Eco-Romanticism of Gary Snyder*.
40 The first occurrence of the word in English is thought to be in 1873. The American George Perkins Marsh had, however, already articulated some of the fundamentals of the "scientific" understanding of ecology in *Man and Nature* (1864): see McKusick, *Green Writing*, 29, 31.
41 See McKusick, *Green Writing*, 19.
42 See Moore, *Ecology and Literature*, 89; Rigby, "Romanticism and Ecocriticism," 66: Hall, *Romantic Ecocriticism: Origin and Legacies*, 4.

43 On German and French developments in ecocriticism, see respectively: Goodbody, "German Ecocriticism: An Overview"; and Posthumus, "Écocritique et 'ecocriticism'."
44 See Radkau, *The Age of Ecology: A Global History*, 12–20.
45 Radkau, *The Age of Ecology*, 24.
46 Audier, *La société écologique et ses ennemis*.
47 *Ecology and the Literature of the British Left*, 4, 7.
48 For a discussion of the political diversity of romanticism, with a typological sketch, see Löwy and Sayre, *Romanticism Against the Tide of Modernity*, Chap. 2.
49 See especially Timothy Morton's influential work, *Ecology Without Nature*.
50 See Economides, *The Ecology of Wonder in Romantic and Postmodern Literature*, introduction.
51 One eco-Marxist, however, has gone a step further and suggested that the term "Anthropocene" should be replaced by "Capitalocene:" see Moore, *Capitalism in the Web of Life*.

Works Cited

Audier, Serge. *La société écologique et ses ennemis: pour une histoire alternative de l'émancipation*. Paris: La Découverte, 2017.

Baschet, Jérôme. *Défaire la tyrannie du présent. Temporalites émergentes et futurs inédits*. Paris: La Découverte, 2018.

de Chateaubriand, François-René. *Le génie du christianisme*. Paris: Garnier Flammarion, 1966.

Clare, John. *Selected Poems*. Edited by Geoffrey Summerfield. London: Penguin, 1990.

Clare, John. *Selected Poetry and Prose*. Edited by Merryn and Raymond Williams. London and New York: Methuen, 1986.

Dickens, Charles. *Hard Times*, New York: Harper and Row, 1965.

The Ecocriticism Reader: Landmarks in Literary Ecology. Edited by Cheryll Glotfelty and Harold Fromm. Athens, GA: University of Georgia Press, 1996.

Ecology and the Literature of the British Left: The Red and the Green. Edited by John Rignall and H. Gustav Klaus. Farnham: Ashgate, 2012.

Economides, Louise. *The Ecology of Wonder in Romantic and Postmodern Literature*. New York: Palgrave Macmillan, 2016.

Fischer, Ernst. *Ursprung und Wesen der Romantik*. Frankfurt: Sendler Verlag, 1986.

Goldmann, Lucien. *Le dieu Caché*. Paris: Gallimard, 1955.

Goodbody, Axel. "German Ecocriticism: An Overview." In *Oxford Handbook of Ecocriticism*, edited by Greg Garrard, 547–559. New York: Oxford University Press, 2014.

Goodridge, John. *John Clare and Community*, Cambridge: Cambridge University Press, 2013.

The Green Studies Reader. Edited by Laurence Coupe. London: Routledge, 2000.

Hall, Dewey. *Romantic Ecocriticism: Origin and Legacies*. Lanham MD: Lexington Books, 2016.

von Humboldt, Alexander. *Views of Nature*. Edited by Stephen T. Jackson and Laura Dassow Walls, Chicago IL: University of Chicago Press, 2014.

von Humboldt, Alexander, and Aimé Bonpland. *Essay on the Geography of Plants*. Edited by Stephen T. Jackson. Chicago IL: University of Chicago Press, 2009.

Löwy, Michael and Robert Sayre. *Romanticism Against the Tide of Modernity*. Durham NC: Duke University Press, 2002.

Löwy, Michael and Robert Sayre. "Romanticism and Capitalism." In *A Companion to European Romanticism*, edited by Michael Ferber, 433–449. London: Blackwell, 2005.

Marx, Karl and Friedrich Engels. *Manifesto of the Communist Party* (1848). In *Collected Works*. Translated by R. Dixon et al. 47 vols. New York: International Publishers, 1975, Vol. 6.

McKusick, James C. *Green Writing: Romanticism and Ecology*. New York: St. Martin's Press, 2000.

Moore, Bryan. *Ecology and Literature: Ecocentric Personification from Antiquity to the Twenty-First Century*. New York: Palgrave Macmillan, 2008.

Moore, Jason. *Capitalism in the Web of Life: Ecology and the Accumulation of Capital*. London: Verso, 2015.

Morton, Timothy. *Ecology Without Nature: Rethinking Environmental Aesthetics*. Cambridge MA: Harvard University Press, 2007.

Novalis. *Werke*. Stuttgart: W. Hädecke Verlag, 1924.

Pinkney, Tony. "Romantic Ecology". In *A Companion to Romanticism*, edited by Duncan Wu, 411–419. Oxford: Blackwell, 1998.

Posthumus, Stéphanie. "Écocritique et 'ecocriticism'. Repenser le personnage écologique." *Figura* 36 (2014): 15–33.

Radkau, Joachim. *The Age of Ecology: A Global History*. Translated by Patrick Camiller. Cambridge: Polity Press, 2014.

Rigby, Kate. "Romanticism and Ecocriticism." In *The Oxford Handbook of Ecocriticism*, edited by Greg Garrard, 60–79. New York: Oxford University Press, 2014.

Rousseau, Jean-Jacques. *Discours sur l'origine et les fondements de l'inégalité parmi les hommes* (1755). Paris: Flammarion, 2008.

Rousseau, Jean-Jacques. *Rêveries du promeneur solitaire* (1778). Paris: Flammarion, 2012.

Rozanis, Stephanos. *I Romantiki Exegersi* (The Romantic Insurrection). Athens: Ekdosis Ilektron, 2001.

Schubert, G. H. "Ansichten von der Nachtseite der Naturwissenschaft" (1808). In *Theorie der Romantik*. Stuttgart: Reclam, 2000.

Tovey, Paige. *The Transatlantic Eco-Romanticism of Gary Snyder*. New York: Palgrave Macmillan, 2013.

Weber, Max. *Gesammelte Aufsätze zur Religionssoziologie II. Hinduismus und Buddhismus*. Tübingen: JCB Mohr, 1921.

Weber, Max. "Science as a Vocation." In *Sociological Writings*. Edited by W. Heydebrand; translated by H. H. Gerth and C. W. Mills. New York: Continuum, 1994.

Wulf, Andrea. *The Invention of Nature: The Adventures of Alexander von Humboldt, the Lost Hero of Science*. London: John Murray, 2015.

1 William Bartram's *Travels* and romantic environmentalism

In a study of Alexander von Humboldt and his influence on nineteenth-century American environmentalism, Aaron Sachs highlights awareness of the "chain of connection" and a passionate subjective communion with nature as central to the thought and sensibility of the German explorer and naturalist. Sachs sees Humboldt as ahead of his time in these respects at the opening of the nineteenth century, and suggests that he may indeed have been "the first ecologist."[1] Although in the introduction to this book we pointed to Humboldt's key role in this respect, it could be argued that William Bartram (1739–1823) has an earlier claim to the title. The work usually referred to simply as Bartram's *Travels* has in recent years become a minor classic in the United States, and its present status, it would seem, owes a great deal to the rise of modern environmentalism and ecology. Since the 1960s there has been widespread recognition of its author as an eighteenth-century precursor of the movement that first developed in America via Henry David Thoreau, Ralph Waldo Emerson, and John Muir, and came to further fruition in the twentieth and twenty-first centuries.[2] At the same time, however, some scholars have questioned the extent to which Bartram was exceptional in relation to his contemporaries, and have pointed to contradictions in his work. In this chapter we will examine the environmental vision of Bartram and its place historically, and attempt to show not only that Bartram's work stands out in its far-reaching ecological thrust, but that this perspective is closely associated with a radical critique of his society, itself unusual in the context of his time and place. We will also argue, against those who have foregrounded Bartram's inconsistencies, that his perspective is remarkably sustained, and that the inconsistencies, such as they are, do not reside in his fundamental vision. That vision is a thoroughly romantic one, in terms of our conception of romanticism.

William Bartram was the son of John Bartram, the renowned Quaker botanist of Philadelphia, who corresponded with, and provided specimens for some of the leading lights of European horticulture, botany and "natural history;" John was named botanist of the king for the North American colonies by George III in 1765. William was an artist and accompanied his father to make sketches on several extensive botanical explorations during his youth. But his *Travels*, published in Philadelphia in 1791, long after the excursion itself, is the narrative of a much longer journey made alone, between 1773 and 1777, in the Carolinas, Georgia, and

East and West Florida, as far west as the lower Mississippi River.[3] The trip was subsidized by the English horticulturist John Fothergill, also a Quaker, to whom Bartram sent both plant samples and sketches. The areas traversed by Bartram were largely what the British termed "wilderness." Claimed by the latter as part of their North American possessions, they were still inhabited and controlled almost entirely by Native American tribes and nations (though British settlement was making progressive inroads in some regions). Bartram was passionately interested in, and attracted to, the Indians he encountered and sojourned with, as he was by the wild – though not unmodified – natural environment in which they lived.

Bartram's writings on the trip went through several stages and versions. He first wrote a two-part report – already in the form of a travel narrative – to his patron in England, sent from the South before the expedition had been completed. This report, which has been published,[4] puts the greatest emphasis on descriptions and enumerations of fauna, flora, and habitats, of the kind that its addressee had instructed the traveller to provide. Even in this document aimed at providing "objective" information, however, Bartram gives way in many places to subjective impressions and effusions, evoking his personal responses to the natural world through which he has been travelling. That dimension is much expanded in the travel account written for publication, which Bartram probably began soon after his return to Philadelphia in 1777. Considerable fragments of a first manuscript version of *Travels* have survived, in which the philosophical and literary elements are paramount.[5] This early rendering then underwent an editing process (clearly with the aid of persons other than Bartram himself), before the work was finally published, fourteen years after the conclusion of the trip.

The first edition of *Travels* was generally better received in Europe than in the United States, and in the remaining years of the eighteenth century the work underwent numerous republications and translations there. The first reviews, and introductions to translations, indicate that the book's main interest was initially seen to be scientific, while the literary-philosophical embellishments on the theme of nature were often criticized.[6] This very aspect, however, soon came to be treasured by many romantic writers who fell under the spell of Bartram's book: Coleridge, Wordsworth, Carlyle, and Chateaubriand, to name only the most prominent. In addition to these European romantics, however, *Travels* also came to be admired by a number of nineteenth-century American writers of similar sensibilities, most notably by Thoreau and Emerson, often considered the first major figures of American environmental concern and nature writing. Thoreau cites Bartram's work in *Walden*, and Emerson, to whom Carlyle recommended *Travels* – though Emerson's journals show that he had already discovered it years earlier – also found inspiration in the book.[7]

The North American context

Important to understanding Bartram's vision is the specificity of the North American setting. The first colonists found what they saw as "wilderness" – lands sparsely populated by peoples whose imprint on them was light – but the "civilizing" of them was rapid at the hands of the newcomers. Annette Kolodny points out that in other parts of the world "civilisation" of land was an ancient, slow evolution, whereas "only in America has the entire process remained within historical memory, giving Americans the unique ability to see themselves as the wilful exploiters of the very land that once promised an escape from such necessities."[8] Within that general context, Bartram lived at a time and in a place that were crucial to the process. For the British colonies in the eighteenth century – most markedly in the second half of it – were fast developing into a thoroughly commercial society in which money and market relations were becoming dominant. This early capitalist civilization was still largely based on agriculture, and its expansionist logic called for the continual extension of land possession and development by the colonists. On his trip Bartram encountered many manifestations of this process.

Other eighteenth-century North American travellers directly participated in it. Among those who crisscrossed the "wilderness" areas were colonizers, settlers moving west, land speculators and surveyors. Another common type of traveler was the fur trader, engaged in a different kind of commercial exploitation of the land. While the first group was engaged in overall appropriation and transformation of territory, the fur trader left the land intact but nonetheless profoundly marked ecosystems by sharply depleting certain animal populations. Travel accounts by persons engaged in such activities naturally reflected those concerns in their representations of the natural surroundings. In *A New Voyage to Carolina* (1709), for instance, the colonizer and land speculator John Lawson repeatedly comments in the text on the settlement and trade potentialities of areas discovered, and notes likely locales on an appended map as well.[9] At the other end of the century, the important fur trader Alexander Mackenzie's account of his exploratory ventures in search of the Northwest Passage treats the land principally as the source of raw materials or as a conduit for commerce, and the Indians he deals with as either potentially helpful agents or as hindrances to his enterprises.[10] Even travelers not directly involved in the exploitive activities mentioned were often influenced in their narratives by these conceptions of land, and by perceived reader expectations.

As for the travelers who were studying "natural history," Roderick Nash notes in his study of American attitudes towards the wilderness that they "occupied a vantage point from which wilderness could be regarded with something other than hostility" – the hostility, that is, of the earlier Puritan view of wild nature as the realm of the demonic, or that of the settlers, who sought to subdue and appropriate it. (Their "hostility" was utilitarian, and often mercenary.) Yet up to the middle of the eighteenth century, according to Nash, natural scientists "shared the dominant point of view." Only then was "a new note sounded in descriptive, scientific writing," one more appreciative of the natural world. Even in the later part of the century, though, "[t]he new attitude coexisted with, rather than replaced, the old," and William Bartram's approach stood out as exceptional.[11] Only in the nineteenth century and beyond did some of the themes and perspectives expressed by Bartram become more widespread among those who travelled in wild natural surroundings or reflected on nature more generally. Even then, in the view of Michael Branch, what was involved was a "minority tradition."[12]

In his study of late eighteenth and early nineteenth-century travel writing Nigel Leask pointedly refuses to treat all travellers of European origin as fitting into a single – "imperialist" – mold, and instead insists on the need to fully particularize and contextualize each traveler being analyzed.[13] We share this approach and will apply it, in cultural as well as socio-historical terms, in the discussion of Bartram that follows.

The worldview of William Bartram

What, then, went into the making of Bartram's worldview? Elements of it certainly came out of the prevailing intellectual and scientific currents of his time: the attitude of the "curious" Enlightenment savant and natural history collector,[14] and the deism which he imbibed through his father and his father's milieu. But it has been convincingly argued that Bartram's point of view is significantly different from prevalent eighteenth-century concepts of nature, strongly influenced by the reasonings of "natural theology," which attempted to prove the existence of a beneficent god through observation of nature. According to Bruce Silver, while Bartram seems unaware of recent "proto-evolutionary theories of the European Enlightenment" and therefore "[writes] lyrically of nature as God's great product," his approach differs crucially from that of the proponents of natural theology. He does not attempt to prove God's benevolent design through observation – a strategy which often led natural theologians to ignore phenomena that did not fit, or to engage in intellectual contortions to make them fit their arguments. Bartram, on the other hand, takes God's omnipresent goodness as a given at the outset, and then concentrates on exploring the workings of the natural world without attempting to explain how God's goodness is manifested in every detail. "He wrote about nature as a poet and sometimes thought about it as a philosopher," Silver concludes, "yet his metaphors and morals never overpower the scientific aims of his odyssey."[15]

That this was possible points to another cultural and spiritual association – clearly the strongest one that went into his make-up, along with the scientific. This was Quakerism, although not the predominant form that it had come to take in eighteenth-century Philadelphia. Frederick Tolles's now-classic study shows how, by the middle of that century, the wealthy Quaker merchants in Philadelphia practiced a moderate, mitigated form of their religion – sometimes associated with a rationalistic deism – one that was compatible with the pursuit of individual profit and accumulation, thereby distancing themselves to a degree from the original ideals of Quaker faith.[16] William Bartram grew up in close proximity to this Quaker elite, but he, like John Woolman, was drawn to a more radical Quakerism closer to that of the founders. That faith involved most centrally the value of egalitarianism, and affirmation of the oneness, or solidarity, of the human and natural communities. Larry R. Clarke has traced more specifically the aspects of original Quaker beliefs that bear on Bartram's view of nature: knowledge of God is experiential, in an intuitive mode, and does not require rational demonstration; it is "apophatic," in that it eschews making positive, intellectual statements about God; and it can be reached through observing nature.[17] These aspects of Quaker belief provide a framework for understanding the coexistence in Bartram of an overriding sense of beneficent divinity in nature, and close observation of phenomena which do not all suggest that in and of themselves.

Generally speaking, Bartram drew from the original Quaker spirit a passionate interest in and respect for the natural world, equally in all its parts, and a defense of its integrity – a tradition of Quakerism that has continued to subsist.[18] Two scholars – Donald Brooks Kelley and Kerry S. Walters – have more particularly demonstrated how Bartram's approach to nature can be related to that of other figures of the radical subcurrent of Quakerism we have referred to, the so-called "weighty" or "tender" Quakers, a group that included Anthony Benezet as well as John Woolman. While seventeenth-century American Quakers had shared, to an extent, Puritan attitudes towards nature, in the eighteenth century this group developed what Walters calls "a uniquely Quaker ecological sensibility."[19] Kelley details some key aspects of these Quaker thinkers' credo. They affirmed that God was the only possessor of the earth, and man only a custodian or steward of it, in a protective relationship that should also extend to animals. Benezet called for a "compact" mode of living similar to that of Indians, and both he and Woolman were critical of the greed for wealth that led to environmental waste.[20] While the affinities of Bartram with these ideas are quite clear, Walters also emphasizes Bartram's difference – and originality – in relation to them, in that his vision was less closely tied to a sectarian Quaker outlook.[21]

Indeed, it has been recognized by most students of his work that there is much in Bartram's vision that goes beyond and stands outside of the Quaker as well as the Enlightenment scientific influences. As Larry Clarke points out, "[his] view of nature as the unspoiled fount and norm of virtue, and of primitive man as more virtuous than civilized, is not part of the Quaker tradition," but is rather "a part of his appeal to the emerging Romantic movement."[22] Bartram himself, in fact, possesses an early romantic sensibility in the fullest sense, if romanticism, as we have claimed, is a revolt against the conditions of bourgeois modernity and nostalgia for pre-modern values and life forms.

For William Bartram, unlike his father, did not fit in commercial Philadelphia or more generally in the early capitalistic society of the British colonies. William's father had wished him to go into business, and set him up as an apprentice to a merchant in the late 1750s, then later, in the 1760s, as a shopkeeper in North Carolina, and finally as a commercial rice and indigo planter in Florida. In all of these ventures William was both unsuccessful and unhappy. When in 1773 he left, alone and against the wishes of his father, on his long expedition in the wilderness of the southeast, his departure appears to have been in large part an escape from the mercantile destiny to which he had been consigned, but for which he had discovered himself totally inapt. His departure was in a real sense a retreat from and rejection of his society. He preferred "wild" nature and the Indians who lived closely integrated into the natural environment.[23]

Before turning to Bartram's account of the trip, we will sketch the broad outlines of his philosophical-religious vision. For this passionate Quaker and romantic, at the most general level all the creatures of the world – plants, animals, and humans – are seen to be manifestations of the divine, and as such to possess equal value. Profoundly linked, they are part of an overarching unity of life forms, one which human beings should be in touch with and embrace. Yet while all created beings are part of a non-hierarchical whole, in Bartram's understanding this does not mean a leveling down to the lowest common denominator. On the contrary, as Pamela Regis has noted, "[t]he general movement is toward elevation – the plants Bartram mentions are animal-like; the animals are humanlike; the savages are not savage at all."[24] Only those who are called "civilized" are devalued, but only because their mode of life constitutes a negation of vital principles.

Among Bartram's unpublished manuscripts there are two fragments that very explicitly develop the implications of these convictions, particularly in socio-cultural terms. They present in condensed form some of the main themes that ran through the first manuscript version of *Travels*, much of which was edited out of the final published version. As such they provide an important philosophical and social background for the travel narrative.[25]

Interestingly, on one of these texts Bartram also superimposed some drawings, and these can be seen to project significant meanings in a pictorial palimpsest. On the lower part of the page we see two images: one of a small house that seems to be beside a river, and the other of a man walking alone among trees, leaning on a stick. Above these two is a third drawing, of a horse, which seems to be galloping, or possibly jumping, since it is oriented upwards. The images all involve the natural world, showing fauna and flora, and human beings in a natural setting, both living and moving (traveling) within it. The drawing of the man walking in woods might be a self-portrait, of Bartram the "philosophical pilgrim," as he repeatedly called himself in the manuscript version of *Travels*.[26]

In the texts themselves Bartram levels a scathing critique at so-called "civilized" society. In this society – in his own, that is – "the more Any Man or Woman, approaches to Honesty and simplicity, the more he is accounted a Fool and he is in the broad Road, and hastening on to Poverty, contempt and Misery." It is a society in which "the Passion of covetousness is One of the most formidable … Enimies. This passion if we give way to it becomes insatiable. It is the parant of contention … and contention begets violence and Warr" (304). The pursuit of wealth, and the work ethic (or "industry," the term Bartram uses) which accompanies it, turn civilized human beings away from the true source of life in nature, leading ultimately to "even a species of suicide. How common is it for Men Whose Aim is for excessive affluance, Riches and fashion able Luxery to ruin their constitution and shorten their Lives, through excessive and constant Labour, fatigue and

Watching" (305). For Bartram, then, the mercantile civilization of which he was a part was a system that was unnatural, and as such deeply damaging from a moral as well as a physical point of view.

The author of these fragments goes on to draw the following conclusions: "Thus it appears I think that we act most rationally and virtuously when our Actions seem to operate from simple instinct [or] apparoach nearest to the maners of the Animal creation." Those nations "who yet remain in the simple state of primitive Nature as our Indians, who had had but little intercourse with White People," are closest to this apparently instinctive life, and therefore to true morality (306). The word "apparently" is important here, since in a number of passages in *Travels* Bartram strongly rejects the notion, held by other whites, that the Indians have no developed social institutions but are instinctive "children of nature." For Bartram they have a structured civilization, but one in consonance with, and integrated into the rest of the living creation.

As for animals, with which he associates the Indians in the passage, it is clear elsewhere that Bartram means to raise their status rather than to lower that of human beings to a debased animality. In other portions of the fragments, the author ironizes about the self-styled "Dignity of Human Nature," an unmerited distinction since mankind behaves as an "Absolute tyrant" (310) towards other animals and "probably would long ere this have destroyd the whole animal creation if his Arms were not withheld, by the Supreme Creatory & preserver ..." (319). Bartram, on the other hand, affirms the "Dignity of Animal Nature" (317), and is convinced that animals express themselves by means of actual languages, and demonstrate rational intelligence (308, 319).[27]

Bartram's *Travels*

In the text of *Travels* itself we find many echoes of this overall perspective, and concrete illustrations of it, although the broad social criticism has been much toned down and there are some passages that apparently contradict or attenuate the radical nature of Bartram's vision. These dissonances have led Thomas Hallock to question the coherency of the Bartramian perspective. He calls *Travels* a "slippery account," which vacillates between different, incompatible positions.[28] Yet the very discrepancy between the various manuscripts on the one hand – both the fragments discussed above and the early draft of the travel account – and *Travels* as published on the other, would seem to provide an argument against the charge of inconsistency. It seems plausible that the differences between the early manuscript draft and *Travels* itself can largely be explained by the editing process that led to the published version. This process appears to have resulted in a kind of censorship, one that was probably partially self-imposed and partially exercised by the external editorial hand. In spite of these complicating elements, though, the tenor of the whole, even in the modified, final publication, is in our view abundantly clear. We will return to the question of its inconsistencies later.

From the very outset of *Travels*, the dominant philosophical theme and tone is exhibited. The travel narrative per se is preceded by an introduction. The opening sentence of this preamble signals to the reader the work's focus on "Nature," a focus asserted to be necessary:

> The attention of the traveller should be particularly turned, in the first place, to the various works of Nature, to mark the distinctions of the climates he may explore, and to offer such useful observations on the different productions as may occur.[29]

Here the accent is on the useful contribution to knowledge, particularly for the botanist and zoologist, as Bartram goes on to specify.

But the short second paragraph announces the philosophical/religious credo: "This world, as a glorious apartment of the boundless palace of the sovereign Creator, is furnished with an infinite variety of animated scenes, inexpressibly beautiful and pleasing" (15). Several pages later we read:

> In every order of nature we perceive a variety of qualities distributed amongst individuals, designed for different purposes and uses; yet it appears evident, that the great Author has impartially distributed his favours to his creatures, so that the attributes of each one seem to be of sufficient importance to manifest the divine and inimitable workmanship (17).

The introduction develops the idea that these qualities and attributes are of the most diverse kinds. The living works of nature, vegetable and animal, impress the observer with their beauty, harmony, awesome mystery, singularity, size, complexity, etc., as well as ingenious usefulness for the organism itself and for others, including man. No individual production of nature will exhibit all or even most of the traits in this eminently subjective list, but each possesses at least one that calls forth wondering admiration. Wonder is indeed one of the dominant notes in *Travels*, projecting what Josephine Herbst calls Bartram's "enchanted vision of the universe."[30]

Beyond this expressed sense of marvel, the introduction, like the work as a whole, points to an ecological awareness of the interconnectedness and unity of living beings. More specifically, Bartram suggests that "[t]he vital principle or efficient cause of motion and action, in the animal and vegetable system, … may be more similar than we generally apprehend" (19), and that the "moral system" of animals may not be so different from or inferior to mankind's as is often supposed (21). One of the several anecdotes from his trip recounted in the introduction exemplifies in experiential terms the author's ecological sensitivity. Bartram happens upon a large spider stalking its bumble bee prey. The text then carefully describes how the spider deftly carries out its successful attack by taking into account the anticipated behavior of not only the bee but also the human being that is present – not as a disembodied scientific observer but as a part of the natural situation. Bartram concludes the anecdote by putting it in the context of the larger food chain. The spider hid under some leaves to devour the bee, but "perhaps before night, became himself the delicious evening repast of a bird or lizard" (25).[31]

Over the course of the following narrative, recurrent passages reinforce and develop the vision announced in the introduction. Early on the journey, while paddling up the Alatamaha River in East Florida, the author rapturously meditates "on the marvellous scenes of primitive nature, as yet unmodified by the hand of man" (65). Some months later, on another river in Florida – the St. John's – Bartram reflects that, in contrast with a temporary traveling companion who seemed only to want to establish himself in a trade and prosper, his own "chief happiness consisted in tracing and admiring the infinite power, majesty, and perfection of the great Almighty Creator" (82), adding, but as if in an afterthought, that it came also from the expectation of providing useful knowledge. Later still, when he has come to the upper reaches of the St. John's, in a very remote area inhabited only by Seminole Indians and the few traders who deal with them, Bartram has an experience of paradisiacal plenitude. Camping with a companion, he feels supremely happy and, echoing an ancient trope of Western culture, associates his state with that of a primitive Golden Age. Recalling one night spent by the campfire, the author exclaims:

> How supremely blessed were our hours at this time! plenty of delicious and healthful food ... with contented minds; under no controul, but what reason and ordinate passions dictated, far removed from the seats of strife. Our situation was like that of the primitive state of man, peaceable, contented and sociable. The simple and necessary calls of nature being satisfied, we were altogether as brethren of one family, strangers to envy, malice, and rapine. (109–110)

In this passage Bartram communes with another human being in the midst of nature. But in some others he feels a similar closeness to animals, as in one occurring much later in his travels, in which he:

> advanced into the strawberry plains to regale on the fragrant, delicious fruit, welcomed by communities of the splendid meleagris, the capricious roebuck, and all the free and happy tribes, which possess and inhabit those prolific fields, who appeared to invite, and joined with me in the participation of the bountiful repast presented to us from the lap of nature (282).

Here Bartram is in Cherokee territory, and a few pages later his account provides another ecstatic description of strawberry fields, this time shared not with deer and other wildlife but with a host of Indian girls (288–290). The account of his and his fellow traveler's encounter with the girls, who are gathering and eating berries, and bathing, is sensuous and erotic without being explicitly lewd, and depends for its effect on the lush natural landscape in which it is set. A note of coy humor at the end points to the men's attraction itself being part of "nature." When, "nature prevailing over reason, we wished at least to have a more active part in their delicious sports," a group of older women appeared and "gave the alarm" (289). The story ends with the sharing of strawberries by all.

As he moves in and out of wilderness areas, and through intermediary zones that have to some extent been changed by European presence, Bartram often explicitly states his preference for wild sites as opposed to those which have been "improved" by settlers. This is the case, for instance, when he tells of a time at which he found himself:

> about sixty miles south of the Alatamaha, passing through an uninhabited wilderness. The sudden transition from rich cultivated settlements, to high pine forests, dark and grassy savannas, forms in my opinion no disagreeable contrasts; and the new objects of observation in the works of nature soon reconcile the surprised imagination to the change (42).

The insistence by the author that this is his opinion seems to indicate that he is contradicting a common, contrary reaction. As has been pointed out by many commentators, Bartram's descriptions of the wilds often draw on the "sublime," a literary form then beginning to come to the fore that celebrates the intensity of unbridled and extreme natural phenomena.[32] But clearly the reader of *Travels* was expected to be likely to respond with the more traditional preference for cultivated, "improved" land.

Another context in which the contrast between land touched or untouched by settlement can become tangible, is when Bartram revisits a spot that has undergone a transformation since his previous visit. On the banks of Lake George, for example, at the head of the St. John's River, Bartram recalled that:

> [a]bout fifteen years ago I visited this place, at which time there were no settlements of white people, but all appeared wild and savage; yet in that uncultivated state it possessed an almost inexpressible air of grandeur, which was now entirely changed (101).

Since the earlier visit it had been cleared, planted with crops, and then abandoned, so that it now was unattractive and desert-like (102).[33]

Overwhelmingly, then, Bartram celebrates the wild nature that he travels through and stops in, preferring it to lands settled by Europeans. This point of view, however, is not entirely unadulterated in *Travels*. On several occasions when near the coast and still in the English zone of habitation, Bartram expresses considerable admiration for the "improvements" made on the estates of some of the commercial planters he stays with en route for outlying territories (37, 77, 85). He at the same time regularly praises the characters of the planters themselves.

Moreover, in several places Bartram uses a kind of rhetorical figure that is common in other British-American travel accounts, one such as we alluded to in John Lawson's narrative. The traveler, in a wild landscape still under Indian control, conjures up in his mind's eye – and in the text – what the land could become if under British control and ownership. The implication is that the change would be desirable, and the function of the passage is also to provide exploitable information for those interested in participating in colonizing ventures. One such instance occurs when Bartram is recounting his passage through Creek and Chactaw country. He comments at one point that:

> [t]he territory lying upon this creek, and the space between it and the river, present every appearance of a delightful and fruitful region in some future day, it being a rich soil, and exceedingly well situated for every branch of agriculture and grazing ...

He adds that waterways from it would allow "uninterrupted navigation to the bay of Mexico and the Atlantic Ocean" (309).

Yet these passages are limited in number, and one of them is revealing because in it Bartram seems to betray his true viewpoint in spite of himself:

> This vast plain, together with the forests contiguous to it, if permitted (by the Siminoles who are sovereigns of these realms) to be in possession and under the culture of industrious planters and mechanics, would in a little time exhibit other scenes than it does at present, delightful as it is; for by the arts of agriculture and commerce, almost every desirable thing in life might be produced and made plentiful here (199).

Two things distinguish this passage from usual formulations in other travel accounts. First, the Seminoles are designated as the "sovereigns" who would have to freely cede their lands if the change were to take place. Second, the phrase "delightful as it is" reveals the underlying regret felt by the author at the thought that the natural beauty of the spot, maintained by the Indians, could disappear. In other passages, in fact, Bartram severely criticizes settlers for the destruction of natural beauties (e.g., 213).

It would seem, then, as we have already suggested, that the dissonant passages, limited as they are, can be laid down either to editorial intervention or to the more subtle pressure exercised by strong prevailing norms on a sensibility that was atypical and marginal to the extreme. The difference in this respect between the early manuscript version and the published work would tend to corroborate the contention that these passages do not correspond to Bartram's deepest sensibilities. As Thomas Hallock himself has noted, "[t]he genetic text shows little interest in expansion ... and portrays nature as a source of inspiration rather than as a commodity."[34] The same can be said of Bartram's being generally well disposed towards the planters near the coast in *Travels*. Christopher Iannini, another commentator who has laid emphasis on Bartram's implication in British colonialism, acknowledges that the figure of the "benevolent planter" does not appear in any known manuscript versions.[35]

Beyond that, Bartram is crucially far more critical of the behavior of whites in the wilderness. Those whom he encountered were settlers, surveyors, and traders, but most often the latter. Although he often traveled alone, and almost always reveled in this mode of travel, he was in many cases obliged – especially due to tensions between Indians and whites – to join caravans, largely constituted of traders, in traversing long stretches of unsettled land. Although he did befriend some traders – particularly among the older and more experienced, whose sagacity he occasionally praises – they were exceptional cases, as he points out himself.

When travelling with traders, Bartram most often chose to stay apart from them on the trail, preferring the society of plants and animals to theirs. Recounting one such expedition with a company of traders crossing some of his most beloved terrain in Florida, Bartram notes that, "[h]aving a good spirited horse under me, I generally kept a-head of my companions, which I often chose to do, as circumstances offered and invited, for the sake of retirement and observation" (187). The mode of travel of these white men is entirely different from that of Indians, and Bartram is highly critical of their practices. They break camp late in the morning, and stop their trek already mid-afternoon. But what outrages Bartram most, as he comments in relation to another, later excursion in Chactaw country further west, is that they press their horses mercilessly with loud cries and curses – accompanied by continual whipping – filling the air with "an incessant uproar and confusion, inexpressibly disagreeable" (351). Such a manner of traveling exhibits the traders' insensitivity to the suffering of animals (Bartram was particularly distressed, on the excursion in Chactaw territory, by that of his own horse, which became seriously exhausted by the caravan's "infernal" pace), and at the same time their total indifference to the natural scenes they pass through.

The cruelty of white travelers in wilderness extends to wild animals as well. Bartram tells of one incident in which he and his traveling companions were approached by an alligator at night. The person who had discovered it was quickly joined by all the others in the party, "for it was a rare piece of sport." They attacked the alligator with firebrands and rammed javelins down his throat. While a few were in favor of "putting an end to his life and sufferings with a rifle ball, … the majority thought this would too soon deprive them of the diversion and pleasure of exercising their various inventions of torture" (210). Only when they grew tired of the game did they put the animal out of its misery. Bartram reports a similar occurrence at a much later date, when another company of traders spied a litter of wolf cubs, chased them and captured one of them: "one of our people caught it by the hind legs, and another beat out its brains with the but [sic] of his gun, – barbarous sport!" (319).

This seemingly widespread indifference to the suffering of animals is often accompanied by a lack of concern with wasting natural resources. Bartram becomes indignant at one point when his fellow travelers kill and cook a tortoise of which they know they will not be able to eat more than half: "my companions, however, seemed regardless, being in the midst of plenty and variety, at any time within our reach, and to be obtained with little or no trouble or fatigue on our part" (159). Even one of the elder traders, one with whom Bartram is on friendly terms, seemed to share this attitude. When a party that the two of them were traveling with sited a herd of deer, "I endeavoured to plead for their lives; but my old friend, though he was a sensible rational and good sort of man, would not yield to my philosophy" (174). These examples strikingly highlight the exceptional nature of Bartram's awareness of ecological issues in the eighteenth-century North American context of seemingly infinite natural abundance.

In contrast to the behavior of the whites stands that of the Indians, whom Bartram perceives to be *at home* in their natural habitats, which they treat in ways that seem to maintain rather than destroy them. These habitats constitute a kind of utopian home for Bartram himself, who is maladapted to, and in flight from the British-American mentality and world. Here, for example, is Bartram's comment on a scene he came upon in East Florida:

> I penetrated the grove, and afterwards entered some almost unlimited savannas and plains, which were absolutely enchanting ; they had been lately burnt by the Indian hunters, and had just now recovered their vernal verdure and gaiety. How happily situated is this retired spot of earth ! What an elysium it is ! where the wandering Siminole, the naked red warrior, roams at large, and after the vigorous chase retires from the scorching heat of the meridian sun ... Seduced by these ... visions of terrestrial happiness, ... I had roved far away ... I turned about, and in the evening regained our camp. (107)

In this passage Bartram displays both an understanding of the way the Indians played an active role in modifying (without dilapidating) the natural environment, and a lyrical sense of their being the inhabitants of nature's utopia.[36]

As reported by Bartram, the Indians he came to know often reciprocally recognized in him a kindred spirit, an exception among Europeans, one who had no desire to appropriate their lands. One Seminole chief gave him "unlimited permission to travel over the country for the purpose of collecting flowers, medicinal plants, &c. saluting me by the name of Puc Puggy, or the Flower hunter, recommending me to the friendship and protection of his people" (163). Another said that "I was as one of his own children or people, and should be protected accordingly" (201). In *Travels*, then, Bartram presents himself as one who found his true home in an unspoiled natural environment, as the adoptive son of its Native American guardians.

The drawings

In the numerous drawings that he produced for his patron John Fothergill during the 1773–1777 trip – a few of which were reproduced in *Travels* – as well as in those he did earlier and later, one finds many aspects of Bartram's idiosyncratic perspective translated into visual terms.[37] Although intended for scientific purposes, they often project Bartram's own way of interpreting and responding to the natural world. As Thomas Hallock has brought out in his study of Bartram's art work, they combine the mimetic, the esthetic and the emotive in a manner that evokes tremendous vitality.[38] Although we cannot here undertake a thorough analysis of the drawings, a few general remarks are in order as a complement to the preceding discussion of Bartram's written works.[39]

One of the key aspects of Bartram's ecological vision – a sense of unity and interconnectedness in the world of living beings – is strongly conveyed in many of the drawings. At the simplest level, and closest to the taxonomic purposes of many of the works, representations of single plants often display the different stages of the organism's life cycle. But beyond this, a large number of images include not one but several life forms, both vegetable and animal, often mixing the two. In one drawing, for instance, we are shown a plant, a bird, a crab, and a seashell, in another several plants, a bird, and a fish. In some cases the organisms are simply juxtaposed, but in others they are shown in interaction. With animals, the most common type of interaction is ingestion. One is pictured in the process of devouring another, as in the textual description of the spider and the bumble bee alluded to earlier. In several drawings we are witness to a snake eating a frog, while in another a larger snake has almost finished swallowing a smaller one. In other images the animal is only looking at its prey, but with clear intent. In one case a bird on a branch looks at a flying insect, in another a water bird directs its beak towards a small crustacean on a beach.

While some of Bartram's representations of life forms are set against a neutral, featureless background, in many of them a further element of interrelatedness is introduced by the depiction of the typical environments of the organisms under display. In the drawing already alluded to of one snake eating another, for instance, the snakes are pictured on the edge of a body of water; in the water is an aquatic plant, and on shore some tufted ground cover and a tortoise. To the various principles of unity at work in Bartram's drawings, which include morphological analogy, life cycle, and food chain, can be added that of habitat.

Although most of the drawings are of non-human subjects, in one a stone calumet head appears alongside a plant, a snail and a crawling insect, establishing a connection with Indian culture.[40] But more broadly, a bridge between human and non-human life is suggested by what has struck many observers as the anthropomorphic expressiveness of many of Bartram's plants and animals. Thomas Hallock cites approvingly, in fact, one of his students who felt that they seem to "have personalities."[41] In this way, as well as through the aesthetic beauty – of line, and sometimes of color – in his imaging of them, Bartram expresses his immense attraction to, and even identification with, the denizens of the natural world.

In recent times the approach that critically analyzes texts in light of the politics of colonial control and domination has often been applied to natural history and nature writing. Mary-Louise Pratt, whose *Imperial Eyes* has been a seminal work for this school of thought,[42] coined the term "anti-conquest" to describe the role played by the early natural-science travelers. From this viewpoint, as Christoph Irmscher has summarized it, "naturalists were the willing, if slightly befuddled, accomplices of Western imperialism,"[43] since their Enlightenment project of imposing and extending rational understanding and classification of nature went hand in hand with the imperial and colonial onslaught. Some analysts of the work of William Bartram have included him within this estimation. James Cox, author of a study of travel literature in the American South, points out that one trend in critical discussion of Bartram and his *Travels* has claimed that "his exploitive tendencies outweigh his 'environmentalist' sensibilities."[44] In what has preceded we hope to have shown the contrary, and moreover that Bartram's true nature is the very opposite of exploitive.

In this chapter we have attempted to demonstrate that Bartram's work represents an early, significant contribution to the development of environmentalism in several respects: in his proto-ecological awareness of biosystems, in his shift of perspective from "anthropocentrism" to "ecocentrism," in his valuing of "wild" nature outside the control of human beings, and in his desire to protect nature from human depredation. We have also shown that Bartram's environmental vision is closely related to his alienation from, and critique of the early capitalist, colonialist society into which he was born, arguing that in spite of some apparent ambiguities and contradictions Bartram's worldview is romantic and radical at its heart.

Many contextual factors can be brought to bear to explain the ways in which Bartram's works sometimes deviate from that core. In eighteenth-century British America, and *a fortiori* in Philadelphia, one of its primary urban centers, the pressure of the dominant ideology – utilitarian, expansionist, technologically manipulative – was intense. In this context, which was also that of a highly volatile, transitional period, many or perhaps even most intellectual productions were rife with unresolved tensions. Hence it is not surprising that – through the editorial process, both internal and external, as we have suggested – Bartram's published work came to manifest some such fault lines. His temperament, which seemingly was tender, and eschewed conflict, also probably played a role, predisposing him to seek reconciliation of positions. In reading *Travels* one does sometimes sense, as Thomas Hallock maintains, a "desire to evade" – to evade confrontation with intractable, unpleasant realities.[45] This is by no means always the case, however. And, most importantly, Bartram's occasional hesitancies should not obscure the nature and power of his essential vision.

Notes

1 Sachs, *The Humboldt Current*, 2.
2 See notably Buell, *The Environmental Imagination*, 62, 77, 81; and Branch, "Indexing American Possibilities," in *The Ecocriticism Reader*, 288, 289. Branch is the editor of *Reading the Roots: Nature Writing Before Walden*. In addition to scholarly

interest connected with modern environmentalism, Bartram has come to be known and fervently admired by a wide public. The two organizations devoted to William Bartram today – the Bartram Trail Conference and the North Carolina Bartram Trail Society – have well-appointed websites publicizing their activities.

3 The full title of the work is *Travels through North and South Carolina, Georgia, East and West Florida, the Cherokee Country, the Extensive Territories of the Muscogulges or Creek Confederacy, and the Country of the Chactaws.*

4 The manuscript was first published as "Travels in Georgia and Florida, 1773–74: A Report to Dr. John Fothergill." More recently it has been republished in Bartram, *Travels and Other Writings.*

5 This text has been transcribed and carefully studied in Nancy E. Hoffmann's doctoral thesis: "The Construction of William Bartram's Narrative Natural History."

6 See Cutting, *John and William Bartram, William Byrd II and St. John de Crèvecoeur*, x, 37, 68.

7 See Thoreau, *A Week on the Concord and Merrimack Rivers; Walden, or Life in the Woods; The Maine Woods; Cape Cod*, 376–377; and Emerson, *Selected Journals 1820–1842*, 487. On the influences of Bartram's *Travels* on both European and American nineteenth-century writers, see Cutting, *John and William Bartram*, 45, 57; Fagin, *William Bartram: Interpreter of the American Landscape*, 128–200. Fagin mentions, in addition to Thoreau and Emerson, Poe's friend and supporter Thomas Holley Chivers as another American admirer of Bartram. See also Slaughter, *The Natures of John and William Bartram*, xvi, 163, 248, 257. Slaughter sees a particularly strong affinity between William Bartram and Thoreau, calling the latter an "heir" to the former's "naturalistic legacy" (248). In their introduction to *William Bartram, The Search for Nature's Design*, Hallock and Hoffmann speak of Bartram having "inspired … an entire tradition of nature writing in America" (2).

8 Kolodny, "Unearthing Herstory," 175.

9 See Lawson, *A New Voyage to Carolina.* Lawson also collected botanical specimens for a European patron, however, and his work contains elements of natural history as well.

10 For a modern edition of the work that follows the first edition of London, 1801, see Mackenzie, *Voyages from Montreal on the River St. Laurence.* For an analysis of Mackenzie's account, see Sayre, "Alexander Mackenzie's Search for the Northwest Passage," 121–137.

11 Nash, *Wilderness and the American Mind*, 53, 55.

12 Branch, "Indexing American Possibilities," 286.

13 See Leask, *Curiosity and the Aesthetics of Travel Writing*, introduction.

14 Pamela Regis, who emphasizes the scientific side of Bartram, takes her cue from a passage in *Travels* to identify "curiosity" as his "attendant spirit": Regis, *Describing Early America*, 64.

15 Silver, "William Bartram's and Other Eighteenth-Century Accounts of Nature," 600, 614.

16 See Tolles, *Meeting House and Counting House: The Quaker Merchants of Colonial Philadelphia.*

17 Clarke, "The Quaker Background of William Bartram's View of Nature," 435–448. In a critical commentary on this article Bruce Silver argues that the ideas Clarke attributes to Quakerism are not specific to it. In making this claim, however, Silver remains at a relatively high level of generality, and does not address all of the particulars adduced by Clarke. See Bruce, "Clarke on the Quaker Background of William Bartram's Approach to Nature," 507–510.

18 See Sox, *Quaker Plant Hunters*, introduction.

19 Walters, "The 'Peaceable Disposition' of Animals," 162. John Gatta uses the term "ecospirituality" with reference to the Quaker ecology of Woolman and Bartram,

in *Making Nature Sacred*, Chap. 2: "Meditating on the Creatures in Early American Life and Letters."

20 See Kelley, "The Evolution of Quaker Theology," 244, 248–249; and Kelley, "Friends and Nature in America: Toward an Eighteenth-Century Quaker Ecology," 261–266.

21 Walters, "The 'Peaceable Disposition' of Animals," 159–160, 162, 165, 167. Walters stresses the influence of Quietism on the eighteenth-century movement of thought, and in another article argues that Bartram's ideas reflect Christian neoplatonism, via the Quakers: see Walters, "The Creator's Boundless Palace: William Bartram's Philosophy of Nature," 309–332.

22 Clarke, "The Quaker Background," 446.

23 On Bartram's life, with emphasis on his relations with his father, see Slaughter, *The Natures of John and William Bartram*. For cogent reflections on his life, see also Bartram, *William Bartram on the Southeastern Indians*, Chap. 1.

24 Regis, *Describing Early America*, 48.

25 "Thoughts on Morality," manuscript fragments without dates, Pennsylvania Historical Society. The texts of these documents have been transcribed in Hoffmann, "The Construction of William Bartram's Narrative Natural History," Appendix C. The page numbers cited for our quotations will refer to this transcription, and will be noted parenthetically in the text.

26 These images have not been mentioned, to our knowledge, in any other discussions of these manuscripts.

27 For further discussion of Bartram's views on animals, see Walters, "The 'Peaceable Disposition' of Animals."

28 Hallock, "'On the Borders of a New World': Ecology, Frontier Plots, and Imperial Elegy in William Bartram's 'Travels'," 125. Also see 112, 120.

29 Bartram, *Travels*, 15. Further references will be made to this edition, and page numbers will be given parenthetically in the text.

30 Herbst, *New Green World*, 119.

31 This is one of the passages often singled out as illustrative of Bartram's ecologically minded elaborations. Another, involving the tiny, short-lived "ephemerae" that inhabit swampy areas, is the example chosen by Chris Magoc for his selection of environmentalist texts, because it "captures the essence of ecological thought in its embryonic period": Magoc, *So Glorious a Landscape: Nature and the Environment in American History and Culture*, 107.

32 Stephanie Volmer points to the wide diversity of styles in *Travels*, including notably the "sublime" and the "picturesque": Volmer, "William Bartram and the Forms of Natural History," 78.

33 In his recent study of the St. Johns River in the eighteenth century and its treatment by Bartram, Daniel Schafer has suggested that the latter often chose not to mention plantations that existed at the time of his 1773–1777 travels, thereby "creat[ing] an idealized world of pristine nature … and based that vision on his observations of the river valley during the mid-1760s": Schafer, *William Bartram and the Ghost Plantations of British East Florida*, 6. As the above-quoted passage evidences, however, the author did in some cases refer to changes that had occurred between the two trips. More generally, as Kathryn Braund points out, in *Travels* Bartram continually refers to transformations of the land, both by European settlers and by Native Americans: Braund, "The Real Worlds of William Bartram's *Travels*," 439.

34 Hallock, "On the Borders of a New World," 112.

35 Iannini, *Fatal Revolutions: Natural History, West Indian Slavery, and the Routes of American Literature*, 199.

36 Since the 1990s a number of scholarly books and articles have questioned the stereotype of the "ecological Indian." While they have been beneficial in

combating oversimplifications, and providing a welcome corrective to the notion that Indians were passive "children of nature" who never actively molded their environments to their purposes, many of the claims made and arguments developed remain speculative and controversial. Many of the issues they have raised are the subject of ongoing debate, and they have not provided any conclusive demonstration that Indian practices, at least before the presence of Europeans began to influence their behavior, produced significantly harmful or self-defeating effects on their environment. For a recent publication which synthesizes past discussions and contributes to the current debate, see *Native Americans and the Environment.*

37 Reproductions from the important Fothergill collection, housed in the British Museum, are to be found in : Bartram, *William Bartram : Botanical and Zoological Drawings, 1756–1788.* A generous selection of drawings and watercolors covering Bartram's entire career is included in Bartram, *Travels and Other Writings.*

38 See Hallock, "Vivification and the Early Art of William Bartram," 43–60.

39 For extended discussions of Bartram's drawings, including their relation to those of other naturalist-illustrators, like Mark Catesby and John James Audubon, see the work of Amy Meyers: Meyers, "Sketches From the Wilderness", PhD dissertation, chapters 3 and 4; and her much later book: Meyers, *Knowing Nature: Art and Science in Philadelphia, 1740–1840,* 137–153.

40 This juxtaposition of an Indian artefact with animal and plant organisms might be interpreted by some as indicative of a naturalization of Native Americans by Bartram, but it should be clear from the previous discussion that such was not the case. Bartram held Indians to be fully "civilized" human beings, but participants in a civilization far more closely integrated within the natural world than that of the European colonizers.

41 Hallock, "Vivification," 53.

42 Pratt, *Imperial Eyes: Travel Writing and Transculturation.*

43 Irmscher, *The Poetics of Natural History: From John Bartram to William James,* 6.

44 Cox, *Traveling South: Travel Narratives and the Construction of American Identity,* 44.

45 Hallock, "On the Borders of a New World," 130.

Works Cited

Bartram, William. "Thoughts on Morality." Bartram Papers, Vol. 1, folders 81, 83. Philadelphia, PA: Pennsylvania Historical Society.

Bartram, William. *Travels.* New York: Penguin, Penguin Nature Library, 1988.

Bartram, William. *Travels and Other Writings.* Edited by Thomas P. Slaughter. New York: The Library of America, 1996.

Bartram, William. "Travels in Georgia and Florida, 1773–74: A Report to Dr. John Fothergill." *Transactions of the American Philosophical Society* 33 (Nov. 1943): 121–242.

Bartram, William. *William Bartram: Botanical and Zoological Drawings, 1756–1788.* Edited by Joseph Ewan. Philadelphia, PA: American Philosophical Society, 1968.

Bartram, William. *William Bartram on the Southeastern Indians.* Edited and annotated by Gregory A. Waselkov and Kathryn E. Holland Braund. Lincoln, NE: University of Nebraska Press, 1995.

Bartram, William. *William Bartram, The Search for Nature's Design: Selected Art, Letters, and Unpublished Writings.* Edited by Thomas Hallock and Nancy E. Hoffmann. Athens GA: University of Georgia Press, 2010.

Branch, Michael. "Indexing American Possibilities: The Natural History Writing of Bartram, Wilson, and *Audubon.*" In *The Ecocriticism Reader: Landmarks in Literary Ecology,* edited by Cheryll Glotfelty and Harold Fromm, 282–302. Athens, GA: University of Georgia Press, 1996.

Branch, Michael, ed. *Reading the Roots: Nature Writing Before Walden*. Athens, GA: University of Georgia Press, 2004.

Braund, Kathryn. "The Real Worlds of William Bartram's *Travels*." In *Bartram's Living Legacy: The Travels and the Nature of the South*, edited by Dorinda G. Dallmeyer, 439–453. Macon, GA: Mercer University Press, 2010.

Buell, Lawrence. *The Environmental Imagination: Thoreau, Nature Writing and the Formation of American Culture*. Cambridge, MA: Harvard University Press, 1995.

Clarke, Larry R. "The Quaker Background of William Bartram's View of Nature." *Journal of the History of Ideas* 46, no. 3 (1985): 435–448.

Cox, John D. *Traveling South: Travel Narratives and the Construction of American Identity*. Athens, GA: University of Georgia Press, 2005.

Cutting, Rose Marie. *John and William Bartram, William Byrd II and St. John de Crèvecoeur: A Reference Guide*. Boston MA: G. K. Hall, 1976.

Emerson, Ralph Waldo. *Selected Journals 1820–1842*. New York: Penguin, Library of America, 2010.

Fagin, Bryllion. *William Bartram: Interpreter of the American Landscape*. Baltimore: Johns Hopkins University Press, 1933.

Gatta, John. *Making Nature Sacred: Literature, Religion and the Environment in America from the Puritans to the Present*. Oxford: Oxford University Press, 2004.

Hallock, Thomas. "'On the Borders of a New World': Ecology, Frontier Plots, and Imperial Elegy in William Bartram's 'Travels'." *South Atlantic Review* 66, no. 4 (Autumn 2001): 109–133.

Hallock, Thomas. "Vivification and the Early Art of William Bartram." In *A Keener Perception: Ecocritical Studies in American Art History*, edited by Alan C. Braddock and Christoph Irmscher. Tuscaloosa: University of Alabama Press, 2009.

Herbst, Josephine. *New Green World*. New York: Hastings House, 1954.

Hoffmann, Nancy E. "The Construction of William Bartram's Narrative *Natural History*: A Genetic Text of the Draft Manuscript for *Travels Through North and South Carolina, Georgia, East and West Florida*." PhD dissertation, University of Pennsylvania, 1996.

Iannini, Christopher P. *Fatal Revolutions: Natural History, West Indian Slavery, and the Routes of American Literature*. Chapel Hill, NC: University of North Carolina Press, for the Omohundro Institute of Early American History and Culture, 2012.

Irmscher, Christoph. *The Poetics of Natural History: From John Bartram to William James*. New Brunswick, NJ: Rutgers University Press, 1999.

Kelley, Donald Brooks. "The Evolution of Quaker Theology and the Unfolding of a Distinctive Quaker Ecological Perspective in Eighteenth-Century America." *Pennsylvania History* 52 (October 1985): 242–253.

Kelley, Donald Brooks. "Friends and Nature in America: Toward an Eighteenth-Century Quaker Ecology." *Pennsylvania History* 53 (October 1986): 257–272.

Kolodny, Annette. "Unearthing Herstory: An Introduction." In *The Ecocriticism Reader: Landmarks in Literary Ecology*, edited by Cheryll Glotfelty and Harold Fromm, 170–181. Athens, GA: University of Georgia Press, 1996.

Lawson, John. *A New Voyage to Carolina*. Edited by Hugh Talmage Lefler. Chapel Hill, NC: University of North Carolina Press, 1967.

Leask, Nigel. *Curiosity and the Aesthetics of Travel Writing, 1770–1840*. Oxford: Oxford University Press, 2002.

Mackenzie, Alexander. *Voyages from Montreal on the River St. Laurence through the Continent of North America to the Frozen and Pacific Oceans in the Years 1789 and 1793, with a preliminary account of the rise, progress, and present state of the fur trade of that country*. Toronto: The Radisson Society of Canada, 1927.

Magoc, Chris J. *So Glorious a Landscape: Nature and the Environment in American History and Culture*. Wilmington DE: American Visions Series, Scholarly Resources Books, 2002.

Meyers, Amy R. W. *Knowing Nature: Art and Science in Philadelphia, 1740–1840*. New Haven, CT: Yale University Press, 2011.

Meyers, Amy R. "Sketches From the Wilderness: Changing Conceptions of Nature in American Natural History Illustration: 1680–1880." PhD dissertation, Yale University, 1985.

Nash, Roderick. *Wilderness and the American Mind*. New Haven, CT: Yale University Press, 3rd edition, 1982.

Native Americans and the Environment: Perspectives on the Ecological Indian. Edited by Michael E. Harkin and David Rich Lewis. Lincoln, NE: University of Nebraska Press, 2007.

Pratt, Mary-Louise. *Imperial Eyes: Travel Writing and Transculturation*. London: Routledge, 1992.

Regis, Pamela. *Describing Early America: Bartram, Jefferson, Crèvecoeur, and the Rhetoric of Natural History*. DeKalb, IL: Northern Illinois University Press, 1992.

Sachs, Aaron. *The Humboldt Current: Nineteenth-Century Exploration and the Roots of American Environmentalism*. New York: Viking, 2006.

Sayre, Robert. "Alexander Mackenzie's Search for the Northwest Passage: the Commercial Imperative (1789–93)." In *The Quest for the Northwest Passage: Knowledge, Nation and Empire, 1576–1806*, edited by Frederic Regard, 121–137. London: Pickering & Chatto, 2013.

Schafer, Daniel. *William Bartram and the Ghost Plantations of British East Florida*. Gainesville, FL: University Press of Florida, 2010.

Silver, Bruce. "Clarke on the Quaker Background of William Bartram's Approach to Nature." *Journal of the History of Ideas* 47, no. 3 (July-Sept. 1986): 507–510.

Silver, Bruce "William Bartram's and Other Eighteenth-Century Accounts of Nature." *Journal of the History of Ideas* 39 (1978): 597–614.

Slaughter, Thomas P. *The Natures of John and William Bartram*. New York: Vintage, 1997.

Sox, David. *Quaker Plant Hunters: From North America's Early Frontier to the South Pacific*. York: Sessions Book Trust, 2004.

Thoreau, Henry David. *A Week on the Concord and Merrimack Rivers; Walden, or Life in the Woods; The Maine Woods; Cape Cod*. New York: Viking Press, Library of America, 1985.

Tolles, Frederick. *Meeting House and Counting House: The Quaker Merchants of Colonial Philadelphia, 1682–1763*. New York: W. W. Norton, 1948.

Volmer, Stephanie. "William Bartram and the Forms of Natural History." In *Fields of Vision: Essays on the Travels of William Bartram*, edited by Kathryn E. Holland Braund and Charlotte M. Porter, 71–80. Tuscaloosa, AL: University of Alabama Press, 2010.

Walters, Kerry S. "The Creator's Boundless Palace: William Bartram's Philosophy of Nature." *Transactions of the Charles S. Peirce Society* 25, no. 3 (Summer 1989): 309–332.

Walters, Kerry S. "The 'Peaceable Disposition' of Animals: William Bartram on the Moral Sensibility of Brute Creation." *Pennsylvania History* 56, no. 3 (July 1989): 157–176.

2 Thomas Cole, painter-prophet of ecological disaster

Thomas Cole (1801–1848) is widely recognized as a major figure in nineteenth-century American art, and as the talented founder of the Hudson River School of landscape painting. But his art has a much deeper and broader cultural, ethical, and social significance. His monumental series *The Course of Empire* (1834–1836) is nothing less than a philosophical, historical and political manifesto, and his landscapes are much more than simple "naturalistic" descriptions. Born in England, Cole emigrated with his family to America in 1818, and established himself in New York in 1825. Although a large part of his paintings depicts American subject matter, his English background is essential to understanding his work.

There is general agreement that Cole is a romantic painter, but one has to ask what is meant by "romanticism" when that is asserted or assumed. For many art historians (as for many literary historians, as discussed in our introduction), it is a period in the history of modern art – roughly the first half of the nineteenth century. Cole is seen as part of this "Romantic epoch," and as influenced by other contemporary romantic artists, in England or the United States, with whom he shares typically romantic traits. We believe that one can better understand Cole's art if one perceives him as part of the larger romantic anti-capitalist cultural movement such as we conceive it. Approaching his work in this way allows us to grasp an underlying coherence of his artistic world, as well as his deeply *adversarial* state of mind.[1] For sure, he was a conservative romantic, but a quite atypical one. His art contains a powerful indictment of the idolatry of Mammon, the madness and destructiveness of war, the *hubris* of warrior commercial empires, the inhumanity of "utilitarian" culture, and the modern destruction of the natural environment. In many respects he seems ahead of his time, and his art resounds with issues and sensibilities of our own time. This is particularly true of his protest against the utilitarian treatment of the environment.[2]

After discussing the social context of Cole's art, and his written work, we will focus on his masterpiece, *The Course of Empire*. There exists a considerable art-historical bibliography on Thomas Cole, and several studies of his paintings are of great value, particularly the work of Alan Wallach and Angela Miller. Since we are not art historians, but interdisciplinary scholars practicing a form of sociology of culture, we have no pretention in this chapter to contribute to the more specifically disciplinary discussions of Cole. But we hope that by interpreting his art in terms of our concept of romantic anti-capitalism, and by highlighting the environmental issue, we can shed some new light on Cole and contribute to a reassessment of his wider cultural significance.

The social context of Cole's art

As has already been suggested, the capitalist system – an unprecedented break with previous societies that instituted the dominance of quantitative market mechanisms over all qualitative social values and relations – first became dominant in Europe and North America in the eighteenth century, most markedly in England and its North-American settler colony. Thomas Cole, then, was raised in and formed by both of the two countries that were in the forefront of burgeoning capitalist modernity. In those countries, the basic structures of the new economic order had already been put into place in the century preceding his birth. As regards British North America, while a thoroughgoing market society progressively developed throughout the eighteenth century the crucial turning point in the "transition to capitalism" was the period around the American Revolution.[3]

Following the War of 1812, however, and especially from the 1820s through the 1840s, American capitalism entered a new phase – a period of "take-off" powered by a number of essential innovations: the rapid growth of manufacturing and of cities, particularly New York, which became the predominant urban center, and the "transportation revolution," constituted by the building and extension of a wide network of "improved" roads, and the development of new forms of transport – the canal, the steamboat, and the railroad. This renovation of modes of transportation facilitated the vast territorial expansion that also characterized the period, involving intensified conflict with indigenous peoples and their forced displacement westward, as well as external imperial conflicts, especially the war with Mexico (1846–1848). All of these developments and others – industrialization, urbanization, aggressive expansion, and commercial exploitation of land – put exceptional pressure on the natural environment in affected areas, often transforming and disfiguring landscapes, and harming eco-systems.[4]

As the United States took this path of rapid socio-economic growth and change in the early nineteenth century – in what is often called "Jacksonian America" – strong ideological currents that supported it became preeminent: optimistic confidence in immanent "progress," defined as material wealth, the doctrine of utilitarianism, and the cult of the "self-made man." Yet at the same time, as recent historiography has increasingly emphasized, there were also undercurrents of anxiety and doubt about the direction being taken by American society, and nostalgic yearnings for an era – partially mythical – in which another ethos had prevailed. These tendencies often were articulated in religious terms. Cole's work resonates with this tendency in American culture of the period, but expresses it in an especially coherent, critical, and "adversarial" form.[5]

One particular social group that expressed this resistance to change, and sense of nostalgia, was what Alan Wallach calls the "New York aristocracy," among whom were the first Americans to recognize Cole's talent and patronize his artistic endeavors. As Wallach points out, however, this group was not an aristocracy in the European sense, but rather one of old wealth, largely commercial, as opposed to the new wealth of the "self-made men." While there is some overlap between Cole's vision and the viewpoint of these American "aristocrats," it is only partial and to a large extent a misleading appearance. They were just as fundamentally attached to mercantile values as the new rich, and their nostalgia for a past order never had the critical depth of Cole's.[6]

In the same way, Cole's stance cannot be identified, politically, with the Whig party backed by many of his patrons from the New York elite of old wealth, against the rising Jacksonian "democrats." Although he supported the Whig William Henry Harrison in the 1840 presidential race, Cole's interest in partisan politics was limited and sporadic.[7] Also, although Whigs and Democrats were opposed on issues like Indian policy and the Mexican War – and in both respects Cole clearly sided with the Whigs – on a more fundamental, socio-economic level they were not divided; both fully espoused the capitalist underpinnings of the social order.

Thus, in spite of some limited linkage to ideological tendencies and groups in the United States, Cole remained throughout his life and career there a relatively marginal figure. The Cole scholar Ellwood Parry emphasizes this "eccentric" positioning of Cole in his essay "On the Outside Looking In: Thomas Cole as a Critic of American Culture."[8] Indeed, Cole seems to have remained throughout his life to a considerable extent an outsider in relation to the American society to which he had emigrated. In an important recent reconsideration of Cole, Tim Barringer has argued that the painter's childhood and youth in Lancashire, England, in an environment similar to that of the Brontë sisters – one in which "industrial modernity" was making rapid incursions on the "sublime uninhabited landscape" and in which Luddite resistance to industrialization was rife – was the crucial determinant of his sensibility, extending throughout his adult life in America.[9]

More than with the viewpoint of any American, Cole's – voiced in his writings, to which we turn now – shows affinities with those of some European visitors to early nineteenth-century America, as well as with English romantic writers whose work he knew well. One of Cole's closest friends in America was the poet and journalist William Cullan Bryant, who played a key role in the introduction of English and European romanticism to the United States. Both Cole and Bryant were strongly attracted to Wordsworth, Coleridge, and Byron in particular.[10]

Cole's written work

In addition to painting and drawing, Cole produced a substantial body of writing, in both prose – which took a variety of forms: journals, notebooks, letters, essays, sketches – and poetry. These expressions of self in the medium of words –

considerably more abundant than those of most nineteenth-century landscape artists – allow us a valuable avenue of access to Cole's views and feelings, and beyond that to his overall worldview, in the context of which we can better interpret his visual creations. Although one can find varying perspectives depending on the audience addressed, some ambiguities, and an evolution in certain respects, what is remarkable overall is a consistent vision that runs like a thread throughout the written work.

The heart of this vision is what could be called a romantic counter-narrative to the dominant ideological perspective of "Jacksonian America," and more generally of the era of early capitalist expansion and development. This latter was a "bourgeois" version of Enlightenment thought, that envisioned human history as the triumphant rise, through stages, from a primitive state of impoverishment and underdevelopment, lack of civilization and refinement, "savagery" and "barbarism," to one of overflowing wealth, technical accomplishment, and general wellbeing in the present. In this view, the future was to be simply a further – and spectacular – unfolding of the same, through the extension of "civilization" in the form of commerce, industry, and technology. Cole's narrative of humanity is that of the romantic worldview, and exactly inverts the bourgeois narrative – the present is a catastrophic fall from an earlier state of plenitude. As for the future, it is unknown, but the continuing development of present trends portends apocalyptic disaster.

Cole's prose writings include many forceful and radical criticisms of the American society in which he lived. He saw the present American scene as a process – the arena in which the drama of the invasion of modernity was taking place before his eyes. From a "wilderness" earlier inhabited by indigenous peoples whose imprint on the natural environment was light, the land was rapidly being transformed, destroying its natural harmony and beauty. In 1829, a few years after his arrival in New York, Cole wrote: "Nothing is more disagreeable to me than the sight of lands that are just clearing with its [sic] prostrate trees – black stumps – burnt and deformed. All the native beauty of the forest taken away by the improving man."[11] The axe used by settlers and developers to take down the trees he loved became for him a potent symbol of destruction.[12]

Cole noted the crucial agency of the modern forms of transportation – railroads, canals, and steamboats – in the process as well. His major public statement on the American landscape – "Essay on American Scenery" – was first presented as a New York Lyceum lecture in 1835, and then published the following year. In it Cole celebrated the wild natural sites that still existed in the Hudson Valley and warned of their endangerment by the forces of "progress." Then, in 1840 he proposed as a subject of Lyceum debate the question: "Are railroads and canals favorable or unfavorable to the morality and happiness of the present generation in the United States?" He was suggesting perhaps by the very phrasing of the question that his own answer would be negative.[13] Indeed, when the following year he delivered his "Essay on American Scenery" to the Catskill Lyceum, he appended a passage that sharply censured the developers who had defaced the land near his home (he had moved there from New York in 1836) by building a railroad through it.[14]

In this resistance to the intrusion of a railroad on a site of natural beauty, Cole's was one of a chorus of English romantic voices – in the larger sense in which we are here defining romanticism – throughout the nineteenth century. In 1844 William Wordsworth carried on an unsuccessful campaign to hinder the passage of a rail line through the Lake District, and anti-railroad protests were initiated in the 1870s by groups that included Thomas Carlyle, William Morris, and John Ruskin. The latter's comment on the effects of this encroachment of industrial modernity in *Fors Clavigera* – his series of letters addressed to English workers – was one that Cole might well have written himself: "You cared not for Gods nor grass, but for cash ... You Enterprised a railroad through the valley – you blasted its rocks away ... The valley is gone, and the Gods with it ..."[15]

In his own protest, Cole does not hesitate to reverse the designations of the bourgeois narrative to which we have referred, by calling those responsible for these depredations "barbarians," and, in a letter to his patron Luman Reed, identifies these barbarians of modern times by their primary motivation, as "dollar-godded utilitarians."[16] In an undated passage from one of his journals, moreover, Cole shows exceptional penetration in pinpointing – and indicting – the productivism and acquisitiveness as ends in themselves exhibited by these "utilitarians" in the early United States:

> The lord of creation has forgotten his high privilege he has sunk into the filth of (avarice) ... This world is a paradise but man makes it a purgatory. – He toils that he may produce more with which to toil. He accumulates in order to aggrandize.

The same note contains a moving statement of what Cole sees as his mission as an artist:

> If I can only induce a single soul ... to look on nature with a loving eye I shall be fully recompensed ... If but one can be made to know that there is something else ... worthy of being known beside the state of the stocks or the latest inventions of the steamboat.[17]

This, then, was Cole's present. Intensely alienated from that reality, he often implies in his writing a yearning for a past state in which human beings respected the natural environment and lived in a harmonious relation to it. In the "Essay on American Scenery" he identified the world of nature as a "Garden of Eden," but one from which modern man had excluded himself. Earlier periods of human society had not. In fact, Cole's longing seems to be directed towards two distinct natural and human environments of the pre-modern past: "wilderness," inhabited originally by "savage" tribes, and nature "pacified" by bucolic communities of small peasants or farmers.

Traces of these two epochs that attracted him most subsisted, in some form, on each of the two continents he knew, although wilderness was of course far more present in America. His comments in letters regarding his two European trips (1829–1832, 1841–1842), in which he notably spent extended periods in the Italian countryside, particularly around Florence, suggest that he was divided, strongly feeling the pull of both.[18]

As for the future, Cole at most accepted as unavoidable the continued development of the United States along the lines it was already taking. In the "Essay on American Scenery" he averred that such was "the road society has to travel," but his ambiguous judgment that "it *may* lead to refinement in the end" (emphasis ours) betrays a deep scepticism about the possibility of America transforming its present crude materialism into a more humane culture.[19] Far more often in Cole's writings – and especially in the unpublished ones in which he could express himself more freely, unhindered by considerations of whom he was addressing – we find premonitions of a nightmarish culmination of the trends he saw in the making. One particularly striking text is an essay/sketch, never published by Cole, entitled "Verdura, or a Tale of After Time." Presented as a vision of the future, it opens:

> The close of the twentieth century was a fearful season in the history of man – for the world was filled with outrage and wrong. The vast continent of America was peopled with its hundreds of millions. The plough had furrowed the last prairie ...[20]

In this imagined dystopia, wild nature has been entirely destroyed, and the principle of unbridled individualism has led to anarchic violence in an unequal struggle for scarce resources. With its emphasis on overpopulation, echoes of Thomas Malthus are doubtless present here, but only in combination with the focus on acquisitiveness that Cole perceived to be at the heart of modern society. Cole saw and criticized violence in the present as well, as an ugly consequence of the expansive thrust for more land to exploit, and he strongly condemned the 1846 war with Mexico.[21]

These themes are reflected also in Cole's copious poetic production, most of which was not published in his lifetime. Many poems, especially earlier ones, show strong affinities with European and English romanticism.[22] But one poem is of particular relevance to our discussion of Cole's romantic critique of modernity, since it provides a textual parallel to *The Course of Empire*.

In "The Lament of the Forest," published in a prominent New York literary magazine in 1838,[23] the poet, alone in a wild, mountainous spot, hears at twilight the personified voice of the forest. It speaks of the general history of nature, and how it underwent cycles, in the old world, of despoilment by man, the "destroyer," resuscitation when warring empires fell, then further destruction at the hand of man. The American continent was long spared, though, since the natives there loved the wilds. When Europeans arrived, however, "the work of desolation was begun ... And fire and steel performed their office well." The Europeans came on like a "human hurricane," and the lamenting voice of the forest foretells immanent, total ruin:

> Our doom is near: behold from east to west
> The skies are darkened by ascending smoke;
> Each hill and every valley is become
> An altar unto Mammon, and the gods
> of man's idolatry – the victims we.[24]

This vision of catastrophe – with its strongly apocalyptic strains – has parallels with currents of existential apprehension in British romanticism. Blake's bleak portrait of modern devastation – a "desolated earth" increasingly covered with "dark Satanic mills" – portended apocalypse although in the ultimately salvational framework of his religious belief. Lord Byron's poem "Darkness" (1816) goes further, in imagining the ultimate death of nature without divine redemption. And Mary Shelley's novel *The Last Man* (1826) pictures the end of humanity due to a plague spreading through warring, "civilized" nations.[25] In all of these projections of apocalyptic angst there is at least an implicit recognition of human responsibility for natural destruction, but what is striking in Cole's poem, and in the prose fantasy "Verdura" previously cited, is that the linkage is direct. In the poem, moreover, it is nature itself, in the form of the forest, that speaks. As we will see, *The Course of Empire* expresses Cole's foreboding vision of manmade ruin in a series of paintings, fusing his historical vision of old and new worlds into a single pictorial progression.

The Course of Empire *as a romantic vision*

The series under the title *The Course of Empire* (1834–1836) is widely seen as Cole's most significant work of art, his masterpiece. Intended for the gallery in the mansion of his patron Luman Reed, it is composed of five canvases ordered in an historical sequence: I. The Savage State; II. The Arcadian or Pastoral State; III. The Consummation of Empire; IV. Destruction; and V. Desolation. The paintings all portray the same scene – a large bay surrounded by hills – but at different times of day, running from dawn to dusk, which parallel the different stages of human civilization pictured in their progression. In his advertisement for the exhibition of the series, Cole quotes two lines from Canto IV of Byron's *Childe Harold*: "First freedom and then Glory; when that fails, Wealth, vice, corruption."[26] The reference is suggestive, and provides a key to the first three paintings of the series.

The Course of Empire was widely popular with the American public, but this popularity seems largely to have been grounded in misunderstanding. Cole's admirers often thought that the work was an allegory of the destiny of European empires – and in particular of the decline of the Roman Empire – with no relation to the United States of his time. And most of those who seemed to understand that Cole was also speaking of the future of America did not agree with him. For instance, an article published in the weekly *New York Mirror* at the time of its exhibition commented that the cyclical view of history manifested in the series describes "that which has been in all past times," but doesn't apply to the United States, which will be saved from destruction by democracy and material progress.[27] Of course the classical architecture that appears in the series does seem European, but this in itself is not conclusive since the American elite copied Greco-Roman styles in its public buildings. It could similarly be pointed out that the imperial eagle, which appears in some of the paintings, was a symbol of both the Roman and the American empires. As Angela Miller emphasizes:

> The collective denial of Cole's real subject – America in the 1830s – was part of what a later historian termed "a confession and avoidance syndrome" in which Americans ritually repeated the lessons of Old World history only to deny their present applicability ... Couching his national allegory in ancient Roman garb, Cole reminded Americans that the imperial analogy they drew between themselves and the second great republic of the Old World was double-edged.[28]

Alan Wallach has argued convincingly that in *The Course of Empire* Cole set forth in allegorical form a pessimistic philosophy of history and an agrarian republican critique of President Jackson's politics.[29] In our view, however, the main focus of Cole's rejection of the Jacksonian American Empire is a strongly articulated romantic anti-capitalist critique of commercialism linked to imperial ambitions and war.[30] We would also claim that this critique does not come exclusively from an "agrarian republican" perspective, but also one that strongly identifies with "wilderness." As we will try to show through discussion of each panel, the whole cycle is a striking pictorial expression of a romantic vision of history in general, and of the dangers threatening America in particular. It is a "prophetic" work, not in the usual meaning of predicting the future, but in the Biblical sense – a warning of impending catastrophe, unless the people, or the nation, changes its course. Ellwood Parry calls it a jeremiad, warning a heedless people of reversals and disasters ahead. This is a good characterization, with the proviso that the cycle concerns not only "disaster ahead" but processes already taking place.[31]

When considering the view of history held by Cole, which underlies *The Course of Empire*, the question of its relation to the "cyclical" theory of history must be addressed. At least as far back as Greek antiquity conceptions of history as following cycles have held sway, but in eighteenth-century Europe and America a secularized form of it became particularly influential, notably as epitomized in Edward Gibbon's *History of the Decline and Fall of the Roman Empire* (1776–1788; published in the United States in 1804). Such a viewpoint was also developed in an explicitly generalized form in Volney's *Les Ruines, ou méditations sur les révolutions des empires* (*The Ruins, or A Survey of the Revolutions of Empires*; French original: 1791; English translation: 1796). This conception of historical process as involving the rise of a civilization, its maturity, and finally its demise, developed side by side, and in competition with the progressivist vision we earlier alluded to as being directly contradicted by Cole in his writings. Cole's relation to the cyclical theory, however, is more complex.

There can be no doubt that in his overall conceptualization of *The Course of Empire* Cole was strongly influenced by the cyclical view. His quotation from Byron's *Childe Harold*, which pithily summarizes it, indicates that this schema was foremost in his mind. The connection is clearer still if we consider that the original title Cole thought to give to his series was *The Cycle of Mutation*. Moreover, even after choosing the title that the set of paintings would finally be given, in his advertisements for it Cole used the plural form – *The Course of Empires* – to stress the repetitive, cyclical nature of the "course" being represented.[32] Although the cyclical notion is thus without doubt important for Cole, it is only one component of his view of history, which is ultimately composite and open-ended. Americans in the eighteenth and early nineteenth centuries who adopted the cyclical theory – in particular, as emphasized by Wallach, many of the founding fathers who espoused a form of "classical republican theory"[33] – often combined it with progressivism and American exceptionalism: corrupt European societies had experienced repeated cycles of rise and fall, but the American "empire," because it was founded on better principles, would be different, ushering in a new era of unending progress.[34]

Cole rejected this optimistic, self-congratulatory perspective on America, and warned that there too, a rise could be followed by a fall. But it does not follow that he believed the American cycle would simply be a part of the eternal repetition of history's cyclical pattern. As we saw in the discussion of his writings, there is a pronounced *apocalyptic* strain in Cole's sensibility. In both poetry and prose he expressed the fear that the kinds of depredation made by modern American civilization upon the natural environment might lead to irreparable damage and disaster. In this sense, he opposed to the prevalent American optimism a pessimism that goes beyond that of the cyclical view. For he suggested – notably in the unpublished dystopic sketch "Verdura" – that American civilization might not just fall victim to the same cyclical fate as had European ones, but that it might bring about a desolation from which there was no return. Such a fear, of course, has a particular resonance for us today.

Cole's prognostic, however, was not uniformly deterministic. His public pronouncements tended to be less bleak than the views expressed in his journals, and in the "Essay on American Scenery," for example, he held out the possibility that, in spite of indications to the contrary in current reality, Americans might finally succeed in appreciating and preserving the "Eden" in which they lived. For him, indeed, a crucial function of art – especially landscape art – was to further that goal. Although the prospects for America were grim, the outcome was not decided in advance. This is why we would argue that *The Course of Empire* may be read as a kind of Biblical prophecy – a dire warning and a call to awareness.

Figure 2.1 Thomas Cole, *The Course of Empire: The Savage State*. 1834. Collection of The New York Historical Society.

Figure 2.2 Thomas Cole, *The Course of Empire: The Arcadian or Pastoral State*. 1834. Collection of The New York Historical Society.

Figure 2.3 Thomas Cole, *The Course of Empire: The Consummation of Empire*. 1835–1836. Collection of The New York Historical Society.

Figure 2.4 Thomas Cole, *The Course of Empire: Destruction*. 1836. Collection of The New York Historical Society.

Figure 2.5 Thomas Cole, *The Course of Empire: Desolation*. 1836. Collection of The New York Historical Society.

The past (paintings 1 and 2)

Cole first conceived the idea of a series on the rise and fall of an empire in the late 1820s, before he took his first European trip, as is attested by notebook entries at the time. The project then continued to develop during and after Cole's sojourn in the Old World, adding a dimension from his discovery of and reflections on the ruins of Roman antiquity. He executed a first version of the opening painting in the series during his stay in Europe (1829–1832): *A Wild Scene* (1831–1832).[35] The overall structure of the series is clearly indicated in diagrams Cole made showing the way the five paintings were to be laid out in Lumen Reed's gallery. The first two paintings were placed to the left of a fireplace, one on top of the other, as the last two were, symmetrically, to the right. The middle painting, larger than the others, was placed between them, above the fireplace. This centerpiece was the empire in its maturity – *The Consummation of Empire* – while to its left were represented two states of the past, and to its right two states of a projected future.

The first two paintings were set apart, then, as a pair within the overall set of five paintings, mirroring the pair facing them on the other side of the fireplace. These representations of two imagined social configurations of the past – *The Savage State* and *The Arcadian, or Pastoral State* – treat as separate stages of human development and its relation to the natural environment the aspects of wildness and benign cultivation that were often joined in Cole's paintings. Even in these, however, the distinction is not absolute.

If one compares *The Savage State* with *A Wild Scene*, the version of it painted while the overall project was still in an early phase, several differences are apparent. One stands out as particularly important from our point of view: the portrayal of the "savage" human beings present in the two versions. *A Wild Scene* pictures them only as hunters, and isolated ones at that, while *The Savage State* shows other activities, and provides an image of community life. These representations are situated at the far right of the painting. In the water in the lower foreground are two canoe-like boats being propelled by two paddlers each. Above them, on an open plateau, are a circle of tepee-like dwellings and a central fire around which a large number of human figures are gathered, engaged in a game or dance. As for the hunters further to the left of the painting, most are part of a compact group in the center, emphasizing the collective nature of the enterprise.

All the hunters are in elevated, open areas illuminated by the dawning sun, and are moving swiftly towards the right in pursuit of game. It is as if they were bringing with them the growing light, which is rising to the left but already bathing their village on the upland area to the far right. Here, then, is a figuration of "savages" that implicitly contradicts a stereotype current in Cole's time, of primitive peoples as inferior beings shrouded in the darkness of ignorance.[36] They appear in the painting as a part of their natural environment, living in symbiosis with it through hunting, but also as the creators of human artisanry: in their simple but structured village, and in the light boats that not only efficiently extend mobility but also display artistic elegance in their lines.

The "savages" in the painting are thus shown as members of and participants in a dynamic human community, and as such contrast with a number of Cole's individual paintings, in which a single, or at most several contemporary Native Americans look passively at the wilderness around them, seemingly contemplating their demise. *The Savage State*, on the other hand, pictures primitive society in its full bloom. This comparison, though, raises the question of the location of the scene in *The Savage State*. This first painting of the series in fact is the one that most strongly associates the narrative with America. The dwellings and boats of the "savages" resemble Amerindian tepees and canoes; also, there is a large fir tree – a variety more common in north-eastern America than in western Europe – near the middle of the painting; and finally, the sun rising to the left over a vast body of water might suggest a view looking south on the American east coast. Cole seems, then, to want to establish a link with America at the outset. But the identification is not absolute. Cole mixes the visual references, so that the message conveyed cannot be tied only to America or to Europe. A particularly striking indication of this purposeful mixing of cultural signs is to be seen in a detail of the largest hunter, to the left of the painting. He is wielding a bow, a typically Amerindian weapon, but is bearded, which Amerindians never were. Consequently, he joins in one figure the primitive hunter of the recent North American past, and that of European Antiquity.

With *The Arcadian, or Pastoral State* Cole continues the narration of human development before the advent of the modern empire. As *The Savage State* has something in common with *A Wild Scene*, this second in the series has similarities with Cole's later *Dream of Arcadia* (1838). In *The Arcadian, or Pastoral State* hunting has given way to agriculture and animal husbandry, with a plowman on the crest of a hill to the left and a shepherd with his flock on a grassy knoll in the center. The village has grown, and moved down to the water, while canoes are replaced by larger, perhaps seagoing vessels, which retain, however, the stylish lines of the earlier crafts. The activities of music and dance have evolved, to become leisure pastimes carried on in a wooded clearing. At the lower left an elderly man drawing a geometric figure in the earth evokes the beginnings of scientific thought, and a woman facing him at the lower right, holding the tools of weaving, embodies the practical arts. Architecture has developed as well, as manifested by a large, Stonehenge-like temple that stands on the plateau where once was situated the primitive village. In spite of these evolutions in human society, however, the latter is still pictured as fully integrated into the natural world. Although much of the landscape has been tamed to human habitation and use, it retains its original contours and much of the vegetation, particularly trees. The wilder face of nature is also preserved, visible in the background in the jutting promontory that appears, in different guises, throughout the series, as well as in the majestic peak beyond it, hidden by clouds in *The Savage State* but now discernible in the far distance.

The Arcadian, or Pastoral State includes, however, two details that are premonitions of things to come. In the lower center, the shadowy figure of a helmeted soldier strides towards the right. In his path is a young boy, drawing an image of a soldier on the surface of a stone bridge, as Cole himself specified in his description of the painting. A second, and equally ominous detail appears at the far lower right, partially cut off by the edge of the picture: the jagged trunk of a very large tree. These details prepare the viewer for what is to follow, and suggest that paradise is already endangered.

The modern Empire: corruption, war, desolation (paintings 3, 4, 5)

The last three paintings of the cycle relate to the Empire: its apotheosis, destruction, and the remaining ruins. In a letter to his patron Lumen Reed on September 18, 1833, Cole described the whole set in this way: "man in his progress from barbarism to civilization – to luxury – the vicious state, or state of destruction – and to the state of ruin and desolation."[37] For this Protestant artist, "luxury" and "vice" are inseparable, and lead to ruin. This section of the cycle was probably influenced by Gibbon's and Volney's works, previously mentioned.

The third and middle panel, *The Consummation of Empire*, represents the summit of urban and commercial prosperity. But it is far from a celebratory picture. It is in fact a visual representation of the characteristics of the modern (capitalist) society abhorred by Cole: excessive luxury, hubris, conspicuous wealth, insolent power. Frenetic multitudes swarm around, while a pompous procession follows a Cesar-like figure. In the right foreground is prominently displayed a large, bronze cornucopia, symbol of unlimited opulence. Everything seems *excessive* in the disturbing accumulation of details which makes up the composition of *Consummation*: too many people, too many boats in the river, too many buildings and temples, too much splendor and paraphernalia, too heavy decorations.

Moreover, while there is no clearly designated Temple of Mammon in the painting, there is a suggestive similarity between the monumental building that dominates the left-hand side of the picture and an explicit Temple of Mammon in a later work. As Ellwood Parry, who has studied in detail Cole's unfinished last series *The Cross and the World*, points out, in the study for *The Pilgrim of the World on his Journey*:

> the position of the Temple of Mammon beside the water in the middle-ground offers an interesting parallel with the Parthenon like structure in *The Consummation of Empire* ... On the roof of this building with strange spiral columns, he placed a wheel of fortune and what may be a statue with a cornucopia from which gold coins cascade to the pavement below, much to the delight of the gathered throngs who worship money.[38]

Scholars have compared this painting to J.M.W. Turner's *Dido Building Carthage*, which Cole had earlier seen in London.[39] Indeed there is a similarity in the general structure: a large body of water between high embankments, classic architecture, etc. But the differences are even more visible. The general atmosphere of feverish agitation in Cole is the antithesis of the peace and calm in the Turner painting. Instead of the teeming multitudes of *Consummation*, we see only a few peaceful figures in *Dido Building Carthage*. And above all, while nature is very present in Turner's painting – in the form of a gigantic tree in the forefront – there is no such imposing presence in Cole.

One of the most striking features of *Consummation* is indeed the *almost complete absence of nature*. Little remains except the water and a few naked rocks in the far background. The lush natural environment, so strongly present in paintings 1 and 2, has nearly disappeared under the heavy monumental architecture that extends to the horizon. All that remains of the exuberant flora of the past are some minuscule trees and bushes hidden among the columns of the temple. *Consummation* is the radical negation of landscape and wilderness. Cole intensely disliked the curtailment of nature in the New York City he knew, and perceived that the prosperous metropolises of modern industrial civilization tended implacably to expel the natural world.

Angela Miller has aptly summarized the *ethos* of this painting:

> Natural religion has devolved into a corrupt and idolatrous worship of wealth and power ... Center stage is held by a colossal statue of Minerva, the goddess of culture and wisdom, under whose auspices the

> young mercantile empire has defied natural limits. Two huge beacons, or phari, marking the mouth of the bay commemorate the dominion of commerce … A rigid geometry has been superscribed upon the serpentine lines of the once pastoral landscape.[40]

As in the history of the Roman Empire, the excesses of luxury carry the seeds of corruption, tyranny, and decline. Cole is of course not literally representing New York in this canvas. Although the Roman architecture could be American neo-classic, the population's clothing is ancient and the vessels in the harbor are not steamboats but archaic crafts. However, there seems to be a manifest intent to show the American public what happens when empires grow too rich, powerful, and corrupt. The painting can be seen as a kind of artistic answer to Andrew Jackson's imperial ambitions and unlimited admiration for the Roman Empire.[41]

The fourth painting, *Destruction*, is an impressive landscape of war and murder. On land and sea, soldiers kill each other in confused confrontations, while heavy smoke rises to the sky and strong waves agitate the harbor. The empire being invaded is not a peaceful one. Its symbol is a gigantic statue of a warrior attacking an invisible enemy. The head is missing and one arm is broken, but the whole is a powerful representation of furious military violence. The warriors are ancient – neither Mexican nor American – and the invaders seem to be winning the battle. On the crumbling bridge, a leading figure, on horseback, raises his sword but is unable to prevent the disaster. Equally impotent, another mounted officer gesticulates in the middle of a boat packed full of soldiers. Although the dress and arms of the combatants are antique, the presence of what appears to be a *cannon* defending the immense palace on the right-hand side surely hints that the scene is not only ancient but refers to modern warfare as well.

Many civilians, especially women, are among the victims. Under the colossal statue a slain child lies on the steps, while his mother, attacked by a soldier, raises her arms to the heavens in desperation. At the forefront of the painting a dramatic scene is depicted – a young women, pursued by a warrior, prefers to leap into the void rather than fall into the hands of her persecutor. Cole's sympathy clearly goes to the innocent victims, but the lesson of the painting, if one relates it to the previous one, seems to be that powerful and rich empires wish to expand, become bellicose, provoke wars, and end up being destroyed.

The last panel, *Desolation*, is one of the most powerful. Linda Ferber considers it "perhaps the most original and certainly the most poetic of the five canvasses."[42] Illuminated by the pale light of the moon, the harbor is in ruins; an enormous column is the most visible remains of the prosperous empire. Other ruins and some broken columns can be seen in the background. No human soul is visible in the desolate landscape. But nature has returned. Large trees and all sorts of vegetation cover the place, sometimes growing up over the broken stones and marbles. The bay has now regained its irregular, natural contours. But this is not the nurturing environment of the two first canvases. It is nature after destruction, among ruins, and without any human presence. There is no return to the savage state of painting 1. As we have already suggested, this is not a cyclical story in the sense of a full circle but a pictorial narrative with a beginning and an end.

Ancient ruins are a typical romantic *topos*, and one finds them in many of Cole's paintings, particularly from the periods of his visits to Italy. But this is a different sort of landscape, a sad and desolate world of destruction and waste. There is some similarity with the enigmatic *Titan's Goblet* (1833), in which the strange object seems to be, like the gigantic column in *Desolation*, a prodigious last trace of a disappeared civilization, victim of its hubris. The historical lesson that Cole seems to be aiming at his American public with *The Course of Empire* would thus be that America is rapidly moving from the savage and pastoral stages towards corrupt commercial empire, and that following this path will lead, through war and destruction, to complete desolation.

Cole was a conservative, an enemy of "progress" and a person deeply immersed in a traditionalist and religious culture, but as a romantic and visionary artist he was a very atypical one. As an uncompromising critic of modernity in the aspects we have discussed, he raises issues that are central to radical critiques being made in our own time.[43] Considering the radical and "adversarial" romantic character of Cole's defense of wilderness, it would be a complete misrepresentation to reduce him to being a precursor of the US public parks movement. This is exactly what Frances Dunwell does, however, in an essay with the promising title "Thomas Cole: Inspiration for the Environmental Movement." According to Dunwell, Cole and the Hudson River School "made natural beauty a matter of national pride and identity," and, with the help of people like Andrew Jackson Downing, persuaded countless communities "to establish public parks where all classes of people could come for outdoor recreation and enjoyment of nature." Moreover, Cole and his fellow artists "fueled a national passion for American scenery that led to the preservation of internationally famous landscapes" such as the Yosemite or Grand Canyon parks.[44] Thus Cole's passionate critique of the desolation produced by a modern form of barbarism is reduced to a harmless plea for "outdoor recreation."

It would seem, rather, that in his dramatic protest against the devastation of nature Cole could be considered a forerunner of the present-day ecological movement. Cole was not a lone pioneer, however, but rather part of the broad romantic current in late eighteenth- and nineteenth-century culture that is one of the key intellectual roots of modern ecology and environmentalism.[45] Cole's perspective is of course markedly aesthetic – focused on treasuring landscapes of natural beauty – and rooted in an "enchanted" view of nature that seems far removed from some current ecological concern with climate change, based on strictly scientific forecasts. Moreover, he did not have any political program, and could not envisage any form of action in defense of nature. It would be anachronistic to simply identify him with the present ecological movement, in the United States or elsewhere. But his indignation at the devastations produced by industrial capitalist "progress" has much in common with the spirit of the more radical ecological movements of the twentieth and twenty-first centuries. There are indeed significant romantic components in some modern ecological currents of thought, including a sense of nostalgia for nature as a lost Garden of Eden and admiration for the symbiotic relationship with nature of indigenous communities – notably shared by Naomi Klein, subject of our last chapter – not to mention the ecological perspectives of the Surrealist movement of Chicago, which promoted a poetic and revolutionary celebration of wilderness and the Iroquois Confederation.[46] In this sense Cole – like William Bartram in his own way, several generations earlier – is very much our contemporary.

Notes

1 In her important study of Cole, Angela Miller speaks of his "adversarial romanticism": Miller, *The Empire of the Eye*, Chap. 1, 48.
2 As Alan Wallach emphasizes, the word "utilitarian" was used by Cole "to connote a heartless ethos of capitalist calculation": Wallach, "Thomas Cole's 'River in the Catskills' as Antipastoral," 341.
3 For a summary of the debates around the question, see "The Transition to Capitalism in America: A Panel Discussion," and Merrill, "Putting 'Capitalism' in its Place." See also the important contributions of Henretta, *The Origins of American Capitalism*; and Kulikoff, *The Agrarian Origins of American Capitalism*.
4 See Stokes and Conway, eds., *The Market Revolution in America*, 27.
5 On the various aspects of the period mentioned here, in addition to the previously cited work, see Miller, ed., *The Nature of Jacksonian America*; Hietala, *Manifest Design: Anxious Aggrandizement in Late Jacksonian America*; Sellers, *The Market Revolution: Jacksonian America 1815–1846*; Feller, *The Jacksonian Promise: America, 1815–1840*; Noll, ed., *God and Mammon: Protestants, Money, and the Market, 1790–1860*.
6 Wallach, "Thomas Cole and the Aristocracy," 97–98. On Cole's relations with his patrons, see also Wallach, "Thomas Cole: Landscape and the Course of American Empire."
7 See Miller, "Thomas Cole and Jacksonian America: *The Course of Empire* as Political Allegory," 67.
8 In *Thomas Cole: Drawn to Nature*, exhibition catalog.
9 Barringer, "The Englishness of Thomas Cole," 3.
10 On Cole's connections with English Romanticism, see Harvey, "Reading the 'book of nature': Thomas Cole and the British Romantics;" see also Miller, *The empire of the Eye*, 23–24, 57–58.
11 Cited in Miller, *The Empire of the Eye*, 60.
12 For this image in a broader context, see Cikovsky, "'The Ravages of the Axe': The Meaning of the Tree Stump in Nineteenth-Century American Art."
13 See Baigell, *Thomas Cole*, 24. On Cole's attitude towards railroads, and in particular the one built in the Catskills while he was living there, see Wallach, "'*River in the Catskills*' as Antipastoral."
14 See Schuyler, *Sanctified Landscape*, 44. This later version of the "Essay on American Scenery" is reproduced, under the title "Lecture on American Scenery," in Cole, *The Collected Essays and Prose Sketches of Thomas Cole*. For the added final passage, see 210–212.
15 John Ruskin, "The White Thorn Blossom," *Fors Clavigera: Letters to the Workers and Labourers of Great Britain*, 1886: Letter 8 (May 1871), quoted in Boos, "An Aesthetic

Ecocommunist," 26. On English romantic protests against railroads, see Hess, *William Wordsworth and the Ecology of Authorship*, Chap. 3.

16 See Schuyler, *Sanctified Landscape*, 43. For instances of the term "barbarian," see 37.

17 Cited in Miller, *The Empire of the Eye*, 60. Cole's original "Essay on American Scenery" contains a formulation quite close to the first part of the journal entry: "The spirit of our society is to contrive but not to enjoy – toiling to produce more toil – accumulating in order to aggrandize." See Cole, *Collected Essays and Prose Sketches*, 6.

18 In a letter to one patron, William Dunlap, after his first visit to Italy, Cole wrote that around Florence he had felt set free from "the cares and business of life, the vortex of politics and utilitarianism that is forever whirling at home." There he contrasted Italy with American modernity, but in a letter to another patron, Robert Gilmore, the contrast was with American nature, and Cole confessed that although he loved the Italian scenery and culture he had experienced, nothing there "has affected me so powerfully as ... the wilderness places of America." In a journal entry in 1841, though, during the second trip, Cole regretted that in America "we feel the want of associations such as cling to scenes in the old world." Cited in Baigell, *Thomas Cole*, 18; Ferber, *The Hudson River School*, 168; Miller, *The Empire of the Eye*, 62.

19 Cole, *Collected Essays and Prose Sketches*, 17. Also cited in Wallach, "Thomas Cole: Landscape and the Course of American Empire," 67.

20 Cited in *Thomas Cole: Drawn to Nature*, 82.

21 In one letter Cole referred to it as the "vile Mexican war:" see Miller, *The Empire of the Eye*, 25.

22 See, for example, "Fancy," "The Vision of Life," and "To the Moon:" Cole, *Thomas Cole's Poetry*, 32–37. In addition to the British romantic authors already mentioned, there are perceptible resemblances with the poetry of Shelley and Keats in Cole's as well.

23 Published in *The Knickerbocker*, this version of the poem had been edited by Cole's friend, William Cullen Bryant. It differs only in details of expression, though, from Cole's manuscript version. Both versions are included in Cole, *Thomas Cole's Poetry*, 100–112.

24 Cole, *Thomas Cole's Poetry*, 111–112.

25 On this dystopian, apocalyptic trend in British romanticism, see McKusick, *Green Writing: Romanticism and Ecology*, 29–30; Rigby, "Romanticism and Ecocriticism," 16–18.

26 See Wallach, "Cole, Byron, and the Course of Empire," 378.

27 Cited by Wallach, "Thomas Cole: Landscape and the Course of American Empire," 95.

28 Miller, *The Empire of the Eye*, 34. However, we cannot agree with Miller when she suggests that the series has to do with "economic cycles" or "fluctuations of the market".

29 Wallach, "Thomas Cole: Landscape and the Course of American Empire," 90.

30 In a later essay Wallach argues for the centrality of the related notion of "luxury" in the series: Wallach, "Luxury and the Downfall of Civilization in Thomas Cole's *Course of Empire*," 304–318.

31 Parry, "On the Outside Looking In," 34.

32 See Wallach, "Cole, Byron, and *The Course of Empire*," 376, 378.

33 See Wallach, "Thomas Cole: Landscape and the Course of American Empire," 91.

34 Cole took the title of his series from a stanza in a poem by Bishop Berkeley: "Westward the course of empire takes its way; / The four first Acts already past, / A fifth shall close the Drama with the day; / Time's noblest offspring is the last." As Wallach points out, later in the nineteenth century, after Cole's series appeared, this stanza was increasingly given an entirely optimistic interpretation,

becoming identified with "Manifest Destiny:" Wallach, "Thomas Cole: Landscape and the Course of American Empire," 195, 109–110.

35 See Baigell, *Thomas Cole*, 46.

36 In his dissertation on *The Course of Empire*, Ellwood Parry signals the "pentimento" presence, nearly invisible, of another hunter in the shadowy area over a flat rock near the bottom of the painting. This figure seems to be the trace, almost completely painted out, of an earlier stage of the conception, more like that of *A Wild Scene*, that had not yet developed the symbolic meanings conveyed in *The Savage State*. See Parry, "Thomas Cole's 'The Course of Empire'," 81.

37 Cited in Powell, *Thomas Cole*, 64.

38 Parry, *The Art of Thomas Cole*, 355.

39 Ferber, *The Hudson River School*, 191.

40 Miller, *The Empire of the Eye*, 30.

41 Tim Barringer has also suggested, in conjunction with his contention that England continued to be an essential reference for Cole throughout his life, that *Consummation* "could easily be read as a parodic vision of Imperial London:" Barringer, "The Course of Empires: Landscape and Identity in America and Britain, 1820–1880," 53.

42 Ferber, *The Hudson River School*, 192.

43 An amusing example of recognition by the contemporary Left of Cole's relevance is to be found in the Spring 2016 catalog of the Verso publishing house. The illustration chosen to advertise a book entitled *How Will Capitalism End?*, by W. Streeck, is Cole's painting *Destruction*.

44 Dunwell, "Thomas Cole: Inspiration for the Environmental Movement," 40, 42.

45 On the line of development running from early romanticism to contemporary ecological thought, see the introduction.

46 See Rosemont, "Karl Marx and the Iroquois."

Works Cited

Baigell, Matthew. *Thomas Cole*. New York: Watson-Guptill Publishers, 1981.

Barringer, Tim. "The Course of Empires: Landscape and Identity in America and Britain, 1820–1880." In *American Sublime: Landscape Painting in the United States, 1820–1880*, edited by Andrew Wilton, 38–65. Princeton NJ: Princeton University Press, 2002.

Barringer, Tim. "The Englishness of Thomas Cole." In *The Cultured Canvas: New Perspectives on American Landscape Painting*, edited by Nancy Siegel. Durham, NH: University of New Hampshire Press, 2011.

Boos, Florence S. "An Aesthetic Ecocommunist: Morris the Red and Morris the Green." In *William Morris: Centenary Essays*, edited by Peter Faulkner and Peter Preston, 21–46. Exeter: University of Exeter Press, 1999.

Cikovsky, Nicolai Jr..“'The Ravages of the Axe': The Meaning of the Tree Stump in Nineteenth-Century American Art." *Art Bulletin* 61, no. 4 (1979): 611–626.

Cole, Thomas. *The Collected Essays and Prose Sketches of Thomas Cole*. Edited by Marshall Tymn. St Paul MN: J Colet Press, John Colet Archive of American Literature, no 7, 1980.

Cole, Thomas. *Thomas Cole's Poetry*. Edited by Marshall B. Tymn. York PA: Liberty Cap Books, 1972.

Dunwell, Frances. "Thomas Cole: Inspiration for the Environmental Movement." In *Thomas Cole: Drawn to Nature*. Exhibition catalog. Guest curator Christine Robinson, 37–43. Albany, NY: Albany Institute of History and Art, 1993.

Feller, Daniel. *The Jacksonian Promise: America, 1815–1840*. Baltimore MD: Johns Hopkins University Press, 1995.

Ferber, Linda S. *The Hudson River School: Nature and the American Vision*. New York: Skira Rizzoli; New York Historical Society, 2009.

Harvey, Samantha C. "Reading the 'book of nature': Thomas Cole and the British Romantics." In *Transatlantic Literary Ecologies: Nature and culture in the nineteenth-century Anglophone Atlantic world*, edited by Kevin Hutchings and John Miller, 73–90. London: Routledge, 2017.

Henretta, James. *The Origins of American Capitalism: Collected Essays*. Boston MA: Northeastern University Press, 1991.

Hess, Scott. *William Wordsworth and the Ecology of Authorship: The Roots of Environmentalism in Nineteenth-Century Culture*. Charlottesville, VA: University of Virginia Press, 2012.

Hietala, Thomas R. *Manifest Design: Anxious Aggrandizement in Late Jacksonian America*. Ithaca NY: Cornell University Press, 1985.

Kulikoff, Allan. *The Agrarian Origins of American Capitalism*. Charlottesville, VA: University Press of Virginia, 1992.

McKusick, James C. *Green Writing: Romanticism and Ecology*. New York: St. Martin's Press, 2000.

Merrill, Michael. "Putting 'Capitalism' in its Place: A Review of Recent Literature." *William and Mary Quarterly* 52, no. 2 (1995): 315–326.

Miller, Angela. *The Empire of the Eye: Landscape Representation and American Cultural Politics 1825–1875*. Ithaca NY: Cornell University Press, 1993.

Miller, Angela. "Thomas Cole and Jacksonian America: *The Course of Empire* as Political Allegory." *Prospects* 14 (1989): 65–92.

Miller, Douglas T., ed. *The Nature of Jacksonian America*. New York: John Wiley and Sons, 1972.

Noll, Mark A., ed. *God and Mammon: Protestants, Money, and the Market, 1790–1860*. Oxford: Oxford University Press, 2002.

Parry, Ellwood C. III. *The Art of Thomas Cole: Ambition and Imagination*. Newark, DE: University of Delaware Press, 1988.

Parry, Ellwood C. III. "On the Outside Looking In: Thomas Cole as a Critic of American Culture." In *Thomas Cole: Drawn to Nature*. Exhibition catalog. Guest curator Christine Robinson, 27–36. Albany NY: Albany Institute of History and Art, 1993.

Parry, Ellwood C. III. "Thomas Cole's 'The Course of Empire': A Study in Serial Imagery." PhD dissertation, Yale University, 1970.

Powell, Earl. *Thomas Cole*. New York: H. N. Abrams, 1990.

Rigby, Kate. "Romanticism and Ecocriticism." In *The Oxford Handbook of Ecocriticism*, edited by Greg Garrard, 60–79. Oxford: Oxford University Press, 2014.

Rosemont, Franklin. "Karl Marx and the Iroquois." In *Arsenal. Surrealist Subversion*. Chicago: Black Swann Press, n° 4, 1989.

Schuyler, David. *Sanctified Landscape: Writers, Artists, and the Hudson River Valley, 1820–1909*. Ithaca NY: Cornell University Press, 2012.

Sellers, Charles. *The Market Revolution: Jacksonian America 1815–1846*. Oxford: Oxford University Press, 1991.

Stokes, Melvyn and Stephen Conway, eds. *The Market Revolution in America: Social, Political, and Religious Expressions, 1800–1880*. Charlottesville: University Press of Virginia, 1996.

Thomas Cole: Drawn to Nature. Exhibition catalog. Guest curator Christine Robinson. Albany: Albany Institute of History and Art, 1993.

"The Transition to Capitalism in America: A Panel Discussion." *History Teacher* 27 (1994): 263–288.

Wallach, Alan. "Cole, Byron, and the Course of Empire." *Art Bulletin* 50 (1968): 375–379.

Wallach, Alan. "Luxury and the Downfall of Civilization in Thomas Cole's *Course of Empire*." In *Global Trade and Visual Arts in Federal New England*, edited by Patricia Johnston and Caroline Frank, 304–318. Durham: University of New Hampshire Press, 2014.

Wallach, Alan. "Thomas Cole and the Aristocracy." In *Reading American Art*, edited by Marianne Doezema and Elizabeth Milroy, 79–108. New Haven CT: Yale University Press, 1998.

Wallach, Alan. "Thomas Cole: Landscape and the Course of American Empire." In *Thomas Cole: Landscape into History*, edited by William Truettner and Alan Wallach, 23–111. New Haven CT: Yale University Press, 1994.

Wallach, Alan. "Thomas Cole's 'River in the Catskills' as Antipastoral." *Art Bulletin* 84 (2002): 334–350.

3 William Morris's romantic ecological utopia

In terms of our definition of romanticism as a worldview centered on protest against the capitalist present in the name of values drawn from the pre-capitalist past, William Morris (1834–1896) could be said to be close to an "ideal type" in the sense in which Max Weber construed the term. Morris's critique of capitalism was, in much of his writing, radical and entirely explicit, addressing itself to some of the most fundamental aspects of the capitalist system. As we have seen with the two previously discussed figures, those who express a romantic anti-capitalist revolt do not necessarily understand capitalism in exact economic terms, in terms of the means of production and the class system it institutes, or attack it in the name of its principal economic victims. Morris, however, does in his later career focus very precisely on crucial aspects of capitalism as an economic phenomenon and – also explicitly – links his refusal of its overall organization of life with a sense of the loss of human values incarnated in earlier societies. In that later period he approaches the more specific ideal type of "revolutionary romanticism" as well, in that he clearly articulates his pre-capitalist nostalgia as a "detour" – an inspiration rather than an end in itself – in the quest for a post-capitalist, utopian future. Finally, in relation to the principal concern of this book, throughout his career Morris constitutes one of the most striking examples of the "elective affinity" between romanticism and ecological awareness.

Morris's life and career manifest a tremendous diversity, and yet at the same time a pronounced, underlying unity. He was a prolific author of poetry, prose fiction, translations, essays, and lectures; a sometime painter and architect; a craftsman and designer in a multitude of media (printed, painted, and embroidered textiles, wallpaper, tiles, stained glass, furniture, and artisanally produced books, to mention only some of the more important ones). The breadth of his artistic activities was indeed exceptional. Throughout these multiple forms of expression, though, a singular vision and set of themes prevails. Moreover, the oneness of Morris's vision traverses a lifework that is divided into two clearly distinct periods of approximately equal duration.

The first of these runs roughly from the late 1850s to the late 1870s, and is characterized by Morris's close involvement with the Pre-Raphaelite movement, in association with his friends Edward Burne-Jones and Dante Gabriel Rossetti.

Burne-Jones and Morris first met at Oxford, where both were strongly drawn to the Anglo-Catholic movement, which called for a return to pre-Reformation Christianity and revered the artistry of the Middle Ages; the Pre-Raphaelites carried this medieval inspiration into their overall outlook and their artistic productions. In this period Morris published a collection of poetic romances in a medievalist vein, but also several works that draw on ancient Greek and Icelandic mythology. These all reveal a fascination and yearning for pre-modern cultures, recreated through the imagination, without any political engagement in the present directed toward social change.

In the latter part of the 1870s, however, Morris's thinking shifted towards a realization of the need to act on the realities of the degraded modern world around him, and he became increasingly involved in concrete political activities. These first centered especially on opposition to the neglect, and/or disfigurement through misguided "restoration," of England's historic buildings, leading Morris most notably to create the Society for the Protection of Ancient Buildings ("Anti-Scrape," as he referred to it) in 1877. In the following years Morris's political reflection broadened and became more radical, and in 1883 he declared himself to be a socialist. In 1885 he was one of the founders of the Socialist League, a break-away from the Social-Democratic Federation. Morris later preferred the term "communist" to define his political credo, which he continued to espouse until his death. During this second period of his life, in which his perspective became explicitly revolutionary, Morris developed his reflections in a series of essays and talks. As we will see, although he identified himself as a communist and opposed the anarchist tendencies of his day, his thought, and vision, in fact constitute an original and fruitful blending of libertarian communism and anarchism. Morris was forcefully influenced by Marxism – he studied *Capital* and other works by Marx and Engels, and personally knew and collaborated with the latter as well as with Marx's daughter Eleanor – but his interpretation and use of Marxist thought were very much his own.[1]

The two periods of Morris's creative life, then, are in important respects sharply contrasted: on the one hand, an estheticism that projects longing for past cultural forms through purely literary evocations; on the other, an active political commitment to revolutionary change. And yet throughout Morris's career his vision remains steadfastly romantic, in the sense in which we conceive it. He thus illustrates in a particularly striking way a characteristic of the overall romantic phenomenon: its "hermaphroditic" nature, which joins opposites and spans the full gamut of socio-political stances. As the subtitle – "Romantic to Revolutionary" – of his biography of Morris suggests, E.P. Thompson seems to locate Morris's romanticism in the first period alone. In the framework of our conception of romanticism, however, it constitutes the more fundamental worldview that underlies both.

This persistence of romanticism throughout Morris's life is clearly apparent also if one considers the authors who most influenced and inspired him: Thomas Carlyle, William Cobbett, and most especially John Ruskin. All three manifest strong romantic sensibilities, but each in his own way expresses the romantic impulse in a backward-looking, even reactionary perspective. Yet Morris continued to refer admiringly to them even after he had turned to communism. The influence of Ruskin, who like Morris had been closely associated with the Pre-Raphaelites in the earlier period, was by far the strongest, and Morris acknowledged Ruskin's importance for him many times during his revolutionary period as well as before it.

Interestingly, Ruskin held Morris in reciprocal esteem, not only in the Pre-Raphaelite period but also later. In a letter in 1858 Ruskin comments that he has "seen his [Morris's] poems, just out, about old chivalry, and they are most noble – very, very great indeed – in their own peculiar way." Then twenty-five years later, in 1883, Ruskin refers to a lecture by Morris that has caused controversy, in terms that show his full approval of Morris's forcefully expressed anti-capitalist standpoint:

> The significant change which Mr. Morris made in the title of his recent lecture, from Art and Democracy, to Art and Plutocracy, strikes at the root of the whole matter ... The changes which he so deeply deplored, and so grandly resented, in this once loveliest city [Oxford], are due wholly to the deadly fact that her power is now dependant on the Plutocracy of Knowledge, instead of its Divinity.[2]

Both men, then, were aware of their common ground, which we would identify with the romantic anti-capitalist worldview, even when their political positions were far apart.[3]

A key element of the unity we have been discussing – between the two periods of Morris's life on the one hand, and between Morris and Ruskin on the other – is ecological concern as an important component of the romantic perspective. Similarly to Ruskin – who famously spoke of the "storm cloud of the nineteenth century"[4] – Morris's sensibility throughout his life was marked by a passionate love of nature and an anguished, often angry awareness of its degradation/imperilment in the modern world. All of Morris's other main preoccupations, and themes in his writing, are related to this central matter. Although articulated in very different ways in the Pre-Raphaelite and in the revolutionary periods, and in the various vehicles of expression utilized by Morris in both periods, concern for the natural world and man's relation to it pervades Morris's work as a whole. Most strikingly perhaps, since one rarely finds such a strong connection elsewhere in the literature of revolutionary commitment, in Morris the utopian political and social ideal of the later period is centrally bound up with the natural environment.

In what follows we will discuss the ecological dimension of Morris's work – and of his romanticism – in the major creative forms in which they appear: the artisanry; the romances; the essays and talks; and finally the utopian fantasy considered by many to be Morris's masterpiece, in which his vision achieves its fullest expression: *News from Nowhere*. We will then, in a final section, look at the writers and movements that have been importantly inspired by, and have drawn on, Morris's ecological intuitions.

Morris as craftsman

Hand craftsmanship is often highly valued by romantic artists and authors, since it is a form of production in pre-modern, pre-mechanized societies that focuses principally on quality of workmanship, use value, and often esthetic value as well, rather than commercial gain. Morris not only shared this admiration for well-crafted handiwork, but also made it one of his principal creative outlets. He produced artisanal works of many varieties – usually in collaboration with others – throughout his career, starting in 1861 with the creation of a decorative arts company by himself and others associated with the Pre-Raphaelite Brotherhood. This joint venture, in which Morris played a predominant role from the outset, became in 1875 Morris & Co., owned by himself alone. The company developed from an original focus on stained glass to the production of a wide range of textile and wallpaper designs, becoming well-known and prosperous, and continuing production well beyond Morris's death. (It ceased to exist only in 1940.) This craft activity – which for Morris himself involved mainly design – continued throughout his second, revolutionary period, although at the more intense political moments he somewhat reduced his involvement in the company. It thus constitutes an important element – and a thoroughly romantic one – of continuity in his life work.[5]

Morris & Co. was a business enterprise, and one that at times was highly successful. However, although it was for the most part astutely managed after an amateurish beginning, it was never run for profit maximization. It was rather motivated by the desire to concretize an ideal of craftsmanship that looked back to the artisanal techniques and procedures of the past, in opposition to the current development of cheap, standardized industrial mass production. Morris's commitment to pre-industrial techniques and materials was not rigidly unbending, though, and he did not oppose using more modern approaches when he felt they did not compromise the quality of the final product. In the treatment of textiles he on occasion used machinery and synthetic substances, but he preferred hand work – in block printing, for example – and natural raw materials – most notably vegetable dyes rather than chemical ones.

This preference led to one of two noticeable tensions between Morris's broader framework of ideas and his practices in the area of craft. Artisanal methods and natural substances were generally far more time-consuming and expensive than the newly developed industrial ones, and consequently made it necessary for Morris's company to ask relatively high prices for what it produced, limiting its clientele to the comparatively affluent. Morris was painfully aware that this state of affairs conflicted with his belief that all members of a society should have access to well-made and beautiful accoutrements of their daily lives. While Morris found the dissonance between his general ideal and his practice of artisanal production uncomfortable, from his point of view there was no fundamental contradiction, though, since he held that his social ideals could not be achieved in utopian enclaves within the present society, but only through society-wide revolutionary change.

The second tension went perhaps deeper. For although Morris fervently believed that craft work should express the creativity of the workers who made the artisanal objects, he in fact allowed little freedom of invention to those who implemented his designs, carrying out the actual embroidery, blocking, dyeing, and so forth. (Morris himself only occasionally performed these activities.) In this sense it could be said that he was expressing in his craft designs the individual sensibility of the romantic artist in modern society. There were, nonetheless, collaborative aspects of his work that brought in the creativity of others. In his designs he often drew on medieval precedents that themselves must have involved the inventiveness of craft workers; also, some of his productions were done in conjunction with other artist associates of the company, most especially Burne-Jones. But most fundamentally, Morris would doubtless have contended that only in a future, revolutionized society could worker creativity fully blossom and become generalized.

If we look at Morris's designs, then, not so much as the production of a craft worker per se than as the expression of an individual artist imbued with deep knowledge of craft traditions, we can notice some typically romantic characteristics running throughout. First, when pictorial elements are present, especially in the tapestries and embroideries, they usually represent medieval and antique figures, drawing on mythical and legendary materials. But the thematic element that is paramount in Morris's designs – whether they include human figures or not – is vegetal, and to a lesser extent, animal life forms. Patterns of leaves, branches, stems, flowers, and fruits are the most common, almost omnipresent components of the designs, while full trees, and birds, are present in some. The designs are always well structured, setting up repeating patterns, sometimes as mirror images; yet they usually also project exuberant variety as well. The overall effect most often is to suggest a garden in the English style, or a tended wood, rather than a wilderness. These were the forms of nature to which Morris was most strongly attached.

Romantic nostalgia for pre-modern cultures, and love of nature as against the mechanical and industrial, are thus joined in Morris's design work. Representations of the medieval or legendary past are usually framed in, or integrate within themselves, figures of nature as well. But even when it is not involved as subject matter, medieval culture often informs purely natural motifs due to Morris having been inspired by medieval models. For Morris the past and nature were equally important, and intertwined. In a lecture given in 1884 he thus presents the two crucial aspects in the formation of a designer:

> However original a man may be, he cannot afford to disregard the works of art that have been produced in times past when design was flourishing; he's bound to study old examples, but he is also bound to supplement that by a careful study of nature …".[6]

In neither case, he adds, should the designer slavishly imitate, suggesting thereby that the artist's – and perhaps more generally the human being's – relationship both to the past and to nature should be interactive and creative.

The romances

As was the case with his production of crafts, Morris wrote what he termed "romances" from the very beginning of his creative career to the end of his life, and his output in this area was also extraordinarily prolific. These works, written both in verse and in prose, all draw centrally on the past in some way or other. Indeed, in a lecture delivered in 1889 Morris declared that for him romance was "a power of making the past part of the present."[7] Early on Morris created narratives that drew on medieval chivalric and ancient Greek legendry, and then increasingly – particularly after two visits to Iceland in the early 1870s – turned to Nordic traditions, including Germanic and Scandinavian ones. But his appropriation of these materials was highly personal. His later prose romances, which often conjure up new imagined worlds, have sometimes been called "fantasy novels" and are seen as among the initiators of a genre later exemplified by Tolkien's *Lord of the Rings*.

During Morris's first period, before his coming to political awareness and commitment, the romance offered an imaginary solution to his alienation from the modern world – a retreat from it into an imagined and idealized past. *The Earthly Paradise* (1868–1870), formally reminiscent of *The Canterbury Tales* and set in Chaucerian times, in its prologue exhorts the reader to

> Forget six counties overhung with smoke,
> Forget the snorting steam and piston stroke,
> Forget the spreading of the hideous town ;
> Think rather of the pack-horse on the down,
> And dream of London, small, and white, and clean,
> The clear Thames bordered by its gardens green.[8]

In the "Envoi" that also frames the work, the (modern) narrator of these tales from a happier time identifies himself as "the idle singer of an empty day … Dreamer of dreams, born out of my due time." He then asks rhetorically "Why should I strive to set the crooked straight?" and pointedly refuses the call to "strive to build a shadowy isle of bliss:" "Not the poor singer of an empty day."[9] Working to set the crooked straight is of course precisely the route upon which Morris embarks in his later, socialist period, but he continues at the same time to write romance narratives. In many of these, a metaphorical quest is undertaken to attain a utopian goal.

In the romances of both periods the relation of mankind to nature is crucial. In the early ones, nature, especially in a domesticated form, is the locus of things held most valuable – Guenevere and Launcelot's love in a garden (*The Defence of Guenevere*, 1857), for example – and is often lovingly described as it existed before the depredations of modern industrialism. Some of the early works, though, also articulate the theme of the (modern) narrator's sense of disharmony with, and inability to fully integrate himself into, the natural order due to an anguished awareness of time's passage, and death. The works of the second period, on the other hand, turn away from this tragic perspective of individual angst towards a more active, hopeful, and collective one. The quest entered upon by the hero of these later works involves at the same time achieving harmony with nature and with fellow human beings in community and love, and utopian images of its successful attainment appear in various guises in these narratives. Before reaching these ideal destinations, though, the hero often encounters their opposite: representations of death-in-life that suggest the modern world.[10]

There are, then, significant differences between the first and second period romances in regard to the treatment of the natural world and mankind's relation to it. But several notable constants recur as well. Firstly, throughout them nature in its uncorrupted plenitude appears in representations of some state of the pre-modern past, or in a fantasy world that shares strong similarities with the latter. Often also, nature is identified with that mythical past state of lost paradisiacal innocence – the Garden of Eden. Finally, and crucially, in Morris's romances nature is *magical*, peopled by spirits, witches, spells, and mysterious forces. As in many romantic creations, the world comes to be "reenchanted" through the imagination. To do this, Morris draws on the supernatural elements in ancient, medieval and Nordic legend and myth. In this way he attempts to recreate the *perception* of nature of pre-modern peoples, although this projection of earlier forms of perception is of course filtered through the lens of his personal, romantic sensibility.

Essays and talks

In Morris's second period, from the late 1870s to the end of his life, he published a large number of articles, many of which were the texts of lectures delivered to public audiences, in which he developed his ideas on a range of broad general concerns, from the point of view of the socialist, or "communist," that he had become. The issues covered were diverse, and the various texts often focused more specifically on one or the other of them, but the texts often brought together – seeing them as an interrelated, indissoluble whole – a set of key areas of human life: labor and habitat, social organization and artistic expression, and, crucially, the relation of human beings to nature. These texts also typically included an historical dimension – relating past, present, and future in a configuration that, as we have already suggested, constitutes a nearly ideal-typical expression of revolutionary romanticism.

Near the end of his life, in 1894, Morris gave a talk entitled "How I became a Socialist," which looks back retrospectively on his life and traces his path to socialist commitment. In it he acknowledges Carlyle and Ruskin, especially the latter, as mentors and kindred spirits in his consciousness of revolt against the present state of human society. In a strikingly forceful summing up of his enduring perspective throughout his life, Morris states that "apart from the desire to produce beautiful things, the leading passion of my life has been and is hatred of modern civilization." In the same passage he goes on to epitomize this civilization as a paradox: it mobilizes "stupendous organization – for the misery of life."[11] In earlier texts Morris is no less severe in his indictment of modernity, emphasizing different aspects while seeing them as part of a unified, coherent whole.

At its heart is what Morris often calls "commercialism," which subordinates all other values and considerations to the making of profit. The competition among individuals and enterprises to increase profits leads to massive inequality and the domination of one class by another – systemic traits denounced by socialists in general – but also to effects less often foregrounded by them: the linked degradation of the quality of the products of labor and of the experience of labor. Modern industrial mass production aimed at the generation of profit tends to create inferior products, by using cheap materials and drastically reducing the element of skilled, creative labor in the process. In its mechanized, repetitive form, under conditions aimed at reducing its cost to the bare minimum, labor tends to become a wholly painful, degrading activity – in many cases akin to torture, in fact – one that diminishes rather than fulfills the laborer. The artistry that can bring satisfaction to him is eliminated from the labor process, and art becomes the elite practice of the few, who are themselves often frustrated in a society that does not value esthetic activity.[12]

The human environment created by modern commercial production is also a debased one. New buildings often exhibit the same qualities of cheapness and lack of beauty as other industrial products, and the modernizing tendency leads to ill-conceived attempts to renovate older buildings that destroy their original qualities. But for Morris the environmental harm caused by modernity goes far beyond this. Since the capitalist system is by nature infinitely expansive, it continually expands industrial and commercial towns and cities, chaotically and without regard for beauty or salubriousness. Modern commercial production extends its effects into the countryside, and beyond, into the wilder areas as well. Considering nature to be only a tool at the disposition of human enterprise, it cuts down forests, pollutes bodies of water and vitiates the atmosphere with noxious smoke and gases. Morris denounces these depredations in the strongest possible language, writing of trees "murdered," of rivers turned into "sewers," of the air "choked by filth," and of London itself as a "muck-heap."[13] This ecological devastation is the end result, in the deepest sense, of an approach to nature in modern times that views humans as standing outside nature, aiming to conquer and subjugate it. For Morris, instead of leading to the enrichment of human life this conquest of nature has on the whole impoverished it.[14]

Morris's line of argument thus leads him – and here he follows in the footsteps of Ruskin – to radically reject the definition of "wealth" asserted by the ideologues of capitalism. In his 1885 article, "Useful Work versus Useless Toil," he strikingly summarizes his point of view:

> These things [modern commodities] ... I will for ever refuse to call wealth: they are not wealth, but waste. Wealth is what Nature gives us and what a reasonable man can make out of the gifts of Nature for his reasonable use. The sunlight, the fresh air, the unspoiled face of the earth, food, raiment and housing necessary and decent; the storing up of knowledge of all kinds, and the power of disseminating it, ... works of art, the beauty which man creates when he is most a man ... –all things which serve the pleasures of people, free, manly and uncorrupted. This is wealth.[15]

The year before, in "Art and Socialism," Morris had epitomized the cruel paradox of modern times through a comparison with the classical symbol of deceptive wealth: "(L)ess lucky than King Midas, our green fields and clear waters, nay the very air we breathe are turned not to gold ... but to filth" – a process that Morris insists can only worsen under the "present gospel of Capital."[16]

In sharp contrast to the contemporary state of affairs – in which false wealth devastates true, natural wealth – societies of the past that were extremely different among themselves had in common a mode of coexistence with nature that was not destructive of it as industrial society was. In his essays and talks Morris most often compares medieval with modern society in this respect, suggesting that Gothic buildings and towns blended into the countryside rather than aggressively encroaching on it, embodying "a full sympathy between the works of man and the land they were made for,"[17] while artisanal modes of production left the air and water clean and the forests plentiful. His nostalgia for the medieval heritage of rural England also on occasion takes on a personal note. Like many Romantic writers, Morris recalls an earlier period in his own life when the incursions of modernity had not gone as far as they later did. In a letter to the *Daily Chronicle* in 1895 he evokes the Epping Forest that he had roamed as a boy when it still held bona fide traces of an earlier age, laments its present reduced state, and calls for its restoration.[18]

Beyond the medieval past, however, Morris's writing shows a sensitivity for earlier, tribal societies and the wilder nature that they inhabited and respected. As we have seen, some of the romances celebrate the deeds of Nordic peoples in a harsh, wild landscape, and in one of his last texts Morris favorably compares the austerity of such a life with "our artificial poverty of civilization, so much bitterer for those who suffer under it than the natural poverty of the rudest barbarism."[19] Indeed, although he only mentions Native Americans once in his work, Florence Boos in her excellent study of Morris as an "ecocommunist" cites "his lifelong interest in alleged radical-democratic aspects of pre-medieval and early medieval European tribespeople," and sees significant parallels between Morris's view of nature and society and that of Amerindians.[20]

The Morris of the later period, when he was writing articles and delivering lectures, does not just hark back nostalgically to bygone eras, however, but exhorts his audiences to join with him in working for revolutionary change that would ultimately usher in a utopian society; this society would be entirely new, never before attained in human history, but at the same time would incorporate aspects of the pre-capitalist past that Morris admired. This communistic society would be radically egalitarian, with the elimination of competition for profits, the accumulation of wealth and the domination of one class over another; thus far it resembles that of most socialist and communist thinkers. But Morris's ideal society differs from many others in being informed through and through by his own particular form of the romantic vision.

For Morris beauty, and the art that produces it, would pervade that society, and art would be one and the same with human labor; no longer being produced by an elite, for an elite, it would restore to most work its creative, and therefore fulfilling, nature. Art in this broad sense, which Morris sees as being in accord with natural beauty rather than contrasting with it, would renovate the human environment; it would include both rural and urban planning, and would break down the damaging opposition between city and country that had developed in modernity. Humans would live, and shape their lives, in small-scale communities, exercising a kind of anarchistic, consensual self-government. Most profoundly, Morris aspired, in his ideal society, to a renewed relationship with the natural world; humans would step down from their position of overweening power in relation to nature, learning once again to live in it and as part of it rather than outside, against, and above it. These characteristics of Morris's social and human ideal are concretized in his utopian fantasy, *News from Nowhere*, to which we turn now.[21]

News from Nowhere

In 1889, the year before *News from Nowhere* first appeared in serial form, Morris published a review of another utopian fiction that had recently come out, *Looking Backward* by the American socialist Edward Bellamy. In the review, Morris categorically rejects Bellamy's representation of the ideal socialist society in *Looking Backward*. Arguing that it reflects the thoroughly *modern* temperament of its author, and the perspective of the "professional middle class," Morris criticizes its picture of the good society as a regimented, centralized state organization in which city life is the essential component and for which the countryside and the natural world are only subservient providers. In Bellamy's utopia the human individual is a mere cog, and lives what Morris calls a "machine-life." For Morris, the fact that Bellamy imagines such a utopia to have been achieved without any struggle and strife simply underlines the fact that it does not constitute a fundamental break with commercial capitalism.[22] Morris's own vision of the ideal communist society is in every respect opposed to Bellamy's, and it is clear that *News from Nowhere* was written as a kind of answer to, or anti-*Looking Backward*.

The subtitle of *News from Nowhere*, published in book form in 1891, was "or an epoch of rest: being some chapters from a Utopian Romance;" in addition to emphasizing that the society described comes only after a period of (revolutionary) turmoil, it connects this fiction with Morris's many other romances and suggests its underlying romantic nature as well. Moreover, the phrase "some chapters from" points to its partial, open-ended character; although designated as a "utopia" (a reference to Thomas More's famous work, one of Morris's favorite books), *News from Nowhere* is not intended as a complete blueprint of a future state of society. Rather it is an evocation of what the experience of such a society might be, in the form of a dream – an oniric visit by a late nineteenth-century English time-traveler to the twenty-first century – that the final sentence suggests might be, if others can be inspired by the ideals it represents, "a vision rather than a dream."[23]

As one astute commentator has pointed out, and demonstrated in his analysis, the vision embodied in *News from Nowhere* is of a truly "ecological society."[24] Others have called it an "ecotopia."[25] The narrator, William Guest, explores in his dream the areas in and around London, and up the river Thames, that he had known as polluted, ugly, overcrowded and poverty-stricken in his own time, as they have become in this revolutionized future society. Through direct observation, supplemented by what he hears from inhabitants of this new world that he meets along the way, he discovers that the air and the water are now clean, that the countryside is teeming with animal life and abundant crops, and that many woodlands, including the Epping Forest that he had seen endangered in his time, have been allowed to grow back and are well-tended. To his amazement he sees that the London he knew no longer exists as a single, mammoth metropolis but has become a cluster of smaller communities separated by pleasant, semi-rural landscapes. Gardens and trees are everywhere, and humans live among them in villages, small towns, and country houses open to general occupancy, instead of in industrial agglomerations as before (the narrator is told that Manchester has disappeared altogether).

This renewal of nature, and of the relations of humans to it, has been made possible through the replacement of large-scale capitalist competition, expansion, and exploitation of resources by local, democratic decision-making and cooperation in production for needs (social arrangements akin to those championed by the Russian anarchist Kropotkin). Highly polluting industrial production has been replaced by either artisanal work or – as is suggested without precision of detail – the use of alternative, less polluting forms of energy production (notably the "force-barges," mentioned in chapter 24, that replace steam power).[26] Also, at a fundamental ideological level, the notion – summarized historically in one passage by Clara, the lover of the narrator's guide Dick – of nature as "outside" human beings and a "slave" to them has been replaced by one that sees humans as fully a part of nature.[27] As a result of these far-reaching changes, a destructive dichotomy has been transcended. Hammond, the great-grandfather of the narrator's guide, who is old enough to have been a witness to the mutation, describes it in this way:

> the difference between town and country grew less and less; and it was indeed this world of the country vivified by the thought and briskness of town-bred folk which has produced that happy and leisurely but eager life of which you have had a first taste.[28]

The human habitations in the country are now again, as Morris felt they had been in the Middle Ages, an organic part of the land; as he visits the area around Oxford near the end of his expedition up the Thames, the narrator notices "the little hill of Hinksey, with two or three very pretty stone houses new-grown on it (I use the word advisedly; for they seemed to belong to it) looked down happily on the full streams and waving grass."[29] Much of the land he sees is made up of meadows and fields, used for agriculture, but the narrator reports that Hammond has told him such fields are "everywhere treated as a garden made for the pleasure as well as the livelihood of all."[30] When he earlier had asked Hammond why some forested areas have been kept within this garden-countryside, the old man had explained to him that "we like these pieces of wild nature, and can afford them, so we have them; let alone that as to the forests, we need a great deal of timber …"[31]

The utopian society depicted in *News from Nowhere* is thus animated just as much by the principle of pleasure as of usefulness. From being a source of pain and misery under capitalist conditions, work has now become both pleasure and art, much of the former deriving from the latter. But some of the deepest pleasures come from the relation of its inhabitants to the natural world. This relation is repeatedly described as intensely sensuous, even erotic. Hammond lyrically describes the transformation of feeling that took place after the revolution:

> The spirit of the new days, of our days, was to be delight in the life of the world; intense and overweening love of the very skin and surface of the earth on which man dwells, such as a lover has in the fair flesh of the woman he loves;

And Ellen, the woman the narrator has become enamored with, shares this passion : "O me! O me! How I love the earth, and the seasons, and weather, and all things that deal with it, and all that grows out of it …"[32]

In the utopian world of *News from Nowhere*, then, humans are passionately reunited with nature, as they had been in the pre-modern past, though on a higher level because they no longer have any fear of want. Moreover, a further dimension of this relationship – important in terms of the analysis we have been developing – is introduced in a passage that describes a short exchange between the narrator, Dick and Clara, soon after they have met Ellen. As they come up from the river in the early morning, the three see Ellen in the garden next to the country cottage where they have spent the night, her eyes against the sun "like light jewels," and Dick exclaims:

> Look, guest ... doesn't it all look like one of those very stories out of Grimm that we were talking about up in Bloomsbury? Here we are two lovers wandering about the world, and we have come to a fairy garden, and there is the very fairy herself amidst of it.[33]

Although they subsequently spin out this metaphor half-jokingly, the image is a significant and striking one that might be extended to the narrative as a whole. In his dream – or vision – the narrator has in some sense entered into a world of romance, and seen an enchanted garden.

Morris's influence

As we hope to have demonstrated in this essay, there is a remarkable continuity to Morris's work as a whole, from beginning to end and in all the major genres in which he chose to express himself. Through them he consistently articulated, in various guises, an underlying romantic vision, and in the later part of his career, a revolutionary romantic one. Throughout, the love of the natural world was a strongly dominant note in the diverse expressions of his romantic sensibility, and this feeling developed in his later period into a deepening ecological concern and commitment. The force of Morris's personality, creativity, and activism in projecting this vision, had a tremendous impact on many of his contemporaries, but also on a multitude of individuals and movements coming after him. In concluding, we will look briefly at the main trends and people that drew inspiration from his work, in its combined social and ecological aspects.

The first and most immediate influence was on what came to be called the Garden City movement. Initiated several years after Morris's death, promoted and implemented by a number of English housing reformers, city planners, landscapers, and gardeners, it produced a number of planned towns and neighborhoods in the early twentieth century, especially in the environs of London, drawing to a considerable extent on Morris's ideals and attempting to put them into practice in small-scale communities. The movement subsequently spread to other locations in the British Isles, and internationally.[34] Although Morris's influence on this movement is undeniable, and consequential, it is a paradoxical one in that it is unlikely Morris would have endorsed it had he lived to see it come into being. Although a few were radicals or socialists, many of the participants in the movement were middle-class liberal reformists whose political views did not at all coincide with Morris's. And most importantly, Morris firmly believed that it was in vain to attempt to create pockets of ideal life within the larger capitalist society, and considered utopian communities to be doomed to failure. Thus, as Anna Vaninskaya puts it in a perceptive essay, "To cast [Morris] as the patron saint of the Garden City is as historically justifiable as to call Marx the forefather of modern social democracy."[35]

Beyond its direct influence on the Garden City movement in early twentieth-century Britain, Morris's vision has continued to exert a forceful pull on various thinkers and trends into the later twentieth and the twenty-first centuries. It has been noticed that Morris (along with Ruskin) was one important source in the environmental movement of the 1960s and 1970s – a reference in particular for Charles A. Reich's *The Greening of America.* [36] Different currents of ecological and social thought in the 1980s, 1990s, and beyond have also been affected by his work in one way or another; these would importantly include some of the more radical tendencies – so-called "deep" and "spiritual" ecology, ecofeminism, eco-socialism, and anarcho-socialism.[37] As with the earlier influences, some later environmental trends are closer than others to Morris's overall perspective, and among the radical positions different aspects of Morris are doubtless mobilized depending on specific orientations. But the widespread reverberations of Morris's work and thought, no matter how diversely they may be interpreted, are evident.

We should finally mention, as particularly deep and extensive, the impact of Morris on two late twentieth-century British writers who in their turn have played key roles in radical thought and activism to this day. The first, E.P. Thompson, wrote the monumental biography of Morris that focuses both on his romanticism and his political radicalism. The second, Raymond Williams, was far more interested in Morris's ecological dimension, and he will be the subject of analysis in Chapter 5.

Notes

1 On Morris's life and work, see E.P. Thompson's magisterial biography: *William Morris: Romantic to Revolutionary*. On Morris's relation to anarchism, see Sargent, "William Morris and the Anarchist Tradition," 61–73. On his relation to Marxism, see Paul Thompson, *The Work of William Morris*, 218, 223–224, 250–255, 257–258.

2 Quoted in *William Morris: The Critical Heritage*, 189, 190.

3 On Ruskin's romantic anti-capitalism, see Löwy and Sayre, *Romanticism Against the Tide of Modernity*, 127–146.

4 Ruskin thusly titled his last two lectures, delivered in 1884. They are reproduced in *The Library Edition of the Works of John Ruskin*, vol. 34.

5 On the textiles of Morris and his company, see Fairclough and Leary, *Textiles by William Morris and Morris & Co. 1861–1940*. On his arts and crafts work more generally, see Van Zandt, *The Life and Works of William Morris*. Both works include numerous color reproductions.

6 Quoted in Fairclough and Leary, *Textiles by William Morris and Morris & Co.*, 9.

7 Quoted in Hodgson, *The Romances of William Morris*, 151.

8 Morris, *Selected Writings and Designs*, 68.

9 Morris, *Selected Writings and Designs*, 67–68.

10 For detailed discussion of Morris's romances in relation to the theme of nature, see Hodgson, *The Romances of William Morris*, esp. 64–65, 78–80, 112–114, 143–144, 150–151, 180–186, 190–191. See also Boos, "An Aesthetic Ecocommunist," 23–24.

11 *Political Writings of William Morris*, 243.

12 See "Useful Work and Useless Toil" (1885), in *Political Writings of William Morris*, 90. In this passage Morris includes some scientists, as well as artists, as among the "enthusiasts" who are the "salt" of a "miserable system," while recognizing that these are also sometimes corrupted by commercialism.

13 "The Beauty of Life" (1880), quoted in Boos, "An Aesthetic Ecocommunist," 31; "The Society of the Future" (1887), in *Political Writings of William Morris*, 193;

"Art and Socialism" (1884), in *Political Writings of William Morris*, 116; "The Society of the Future," 196.

14 "Useful Work and Useless Toil" (1885), in *Political Writings of William Morris*, 93–94.

15 "Useful Work and Useless Toil," in *Political Writings of William Morris*, 91.

16 "Art and Socialism" (1884), in *Political Writings of William Morris*, 116.

17 "The Lesser Arts" (1878), in *Political Writings of William Morris*, 46.

18 Morris, *Selected Writings and Designs*, 307–308.

19 Article in *Justice*, May 1, 1896, Morris, *Selected Writings and Designs*, 305.

20 See Boos, "An Aesthetic Ecocommunist," 36–40.

21 On some of these themes in the essays and talks, see especially "The Lesser Arts" (1877), in *Political Writings of William Morris*, 33, 55–56; "Art Under Plutocracy" (1883), in *Political Writings of William Morris*, 58; "The Society of the Future" (1887), in *Political Writings of William Morris*, 201.

22 Morris, *News from Nowhere and Other Writings*, 354, 357.

23 *Three Works by William Morris*, 401.

24 O'Sullivan, "The Ending of the Journey: William Morris, *News from Nowhere* and Ecology," 169.

25 See Morris, *News from Nowhere*, 187, note 57.

26 Concerning the "force-barges," see Morris, *Three Works by William Morris*, 350; and more generally, see O'Sullivan, "The Ending of the Journey," 174, 176.

27 See Morris, *Three Works by William Morris*, 367.

28 Morris, *Three Works by William Morris*, 254.

29 Morris, *Three Works by William Morris*, 374.

30 Morris, *Three Works by William Morris*, 380.

31 Morris, *Three Works by William Morris*, 256.

32 Morris, *Three Works by William Morris*, 316, 391. Morris has been criticized, with some reason, for the tendency in *News from Nowhere* to identify women with nature, childbearing and the providing of food and nurture. The main female characters in the narrative, especially Ellen, do however transcend these stereotypes to a significant extent. In an interesting article that analyzes *News from Nowhere* in ways similar to our treatment of it in other respects, John Bellamy Foster argues that in this work, as elsewhere, Morris expresses what is ultimately a strong feminist perspective: "William Morris's Romantic Revolutionary Ideal," 17–35.

33 Morris, *Three Works by William Morris*, 342.

34 On the first developments of the Garden City movement in relation to Morris, see MacCarthy, *Anarchy and Beauty: William Morris and his Legacy, 1860–1960*, 77–91.

35 Vaninskaya, "William Morris and the Garden City," 136.

36 See Aho, *William Morris, a Reference Guide*, 274, 323, 328.

37 See Boos, "An Aesthetic Ecocommunist," 22, 34, 40–43.

Works Cited

Aho, Gary L. *William Morris, A Reference Guide*. Boston, MA: G.K. Hall, 1985.

Boos, Florence S. "An Aesthetic Ecocommunist: Morris the Red and Morris the Green." In *William Morris: Centenary Essays*, edited by Peter Faulkner and Peter Preston, 21–46. Exeter: University of Exeter Press, 1999.

Fairclough, Oliver and Emmeline Leary. *Textiles by William Morris and Morris & Co. 1861–1940*. London: Thames and Hudson, Birmingham Museums and Art Gallery, 1981.

Foster, John Bellamy. "William Morris's Romantic Revolutionary Ideal: Nature, Labour and Gender in *News from Nowhere*." *The Journal of William Morris Studies* 22, no. 2, (2017): 17–35.

Hodgson, Amanda. *The Romances of William Morris*. Cambridge: Cambridge University Press, 1987.

Löwy, Michael and Robert Sayre. *Romanticism Against the Tide of Modernity*. Durham NC: Duke University Press, 2001.

MacCarthy, Fiona. *Anarchy and Beauty: William Morris and his Legacy, 1860–1960*. New Haven CT: Yale University Press, 2014.

Morris, William. *News from Nowhere and Other Writings*. Edited by Clive Wilmer. Harmondsworth: Penguin, 1993.

Morris, William. *Political Writings of William Morris*. Edited by A. L. Morton. London: Lawrence & Wishart, 1979.

Morris, William. *Selected Writings and Designs*. Edited by Asa Briggs. Harmondsworth: Penguin, 1962.

Morris, William. *Three Works by William Morris*. Edited by A. L. Morton. London: Lawrence & Wishart, 1968.

O'Sullivan, Paddy. "The Ending of the Journey: William Morris, *News from Nowhere* and Ecology." In *William Morris & News from Nowhere: A Vision for Our Time*, edited by Stephen Coleman and Paddy O'Sullivan, 169–181. Bideford: Green Books, 1990.

Ruskin, John. *The Library Edition of the Works of John Ruskin*. Edited by E. T. Cook and Alexander Wedderburn. 39 volumes. London: George Allen, 1905–1912, vol. 34.

Sargent, Lyman Tower. "William Morris and the Anarchist Tradition." In *Socialism and the Literary Artistry of William Morris*, edited by Florence S. Boos and Carole G. Silver, 61–73. Columbia, MO: University of Missouri Press, 1990.

Thompson, E. P. *William Morris: Romantic to Revolutionary*. London: Merlin Press, 1955.

Thompson, Paul. *The Work of William Morris*. London: Quartet Books, 1977.

Vaninskaya, Anna. "William Morris and the Garden City." In *Ecology and the Literature of the British Left: The Red and the Green*, edited by J. Rignall, H. G. Klaus, and V. Cunningham, 125–136. Farnham: Ashgate, 2012.

Van Zandt, Eleanor. *The Life and Works of William Morris*. Bristol: Parragon, 1995.

William Morris: The Critical Heritage. Edited by Peter Faulkner. London: Routledge and Kegan Paul, 1973.

4 Walter Benjamin against the "murder" of nature

Walter Benjamin (1892–1940) is one of the most significant cultural theorists of the twentieth century. A close friend of Gershom Scholem and Theodor W. Adorno, he belongs to an intellectual constellation specific to German-Jewish culture, inspired by Jewish messianism, German romanticism, and modern revolutionary ideas. His writings do not take the form of a theoretical system, but rather deploy an experimental mode of thought in permanent evolution. His first essays are related to early twentieth-century romantic *Zivilisationskritik*, but evidence a strong messianic component and anarchist leanings as well. After Benjamin discovered Marxism in 1924, historical materialism became a decisive constituent of his reflections, but in a heterodox and idiosyncratic form. With the exception of a short period (1933–1935) when he seems to have experimented with "productivism," Benjamin's Marxist writings also display significant romantic and messianic elements in the interpretation of literature, culture, history, and politics. They were thus inseparable dimensions of his thought throughout his career. His last piece, the brief notations known as "Theses on the Philosophy of History," written shortly before his death, is one of the most important documents of modern revolutionary culture. After a failed attempt to escape Nazi-occupied France, Benjamin committed suicide in Portbou, Spain, in September 1940.

Romantic *Zivilisationskritik* and the discovery of Marxism (1913–1930)

Walter Benjamin occupies a unique place in the history of modern revolutionary thought, as the first Marxist to break radically with the ideology of progress. He was also one of the first Marxists to propose a radical critique of the concept of the "exploitation of nature" and of the capitalist destructive – "murderous" is his term – relationship with nature. His thinking therefore has a distinct critical quality that sets it apart from the dominant and "official" forms of historical materialism. This peculiarity has to do with his ability to incorporate into the body of Marxist revolutionary theory insights from the romantic critique of civilization. By his radical deconstruction of the dominant modern view of history as "progress" and his empathy with pre-modern non-destructive attitudes towards nature, Benjamin prepared the ground for some later developments in radical ecological thought.

One of Benjamin's first articles – published in 1913 – is precisely entitled *Romantik*. It is a call for the birth of a "new Romanticism" and an homage to the "romantic desire for beauty, truth and action" that is a supreme moment of modern culture. This inaugural text documents both Benjamin's deep affinity with the romantic tradition – conceived as being simultaneously art, knowledge, and praxis – and a desire to renew it.[1]

Like many Romantic authors, the young Benjamin tends to spiritualize nature, as one can see in the (unpublished by Benjamin) fragment on "color" from 1915: *The Rainbow. A Dialogue on Imagination*. Imagination is "the color of nature, of the mountains, trees, rivers and valleys, but above all of flowers and butterflies, of the seas and the clouds;" and the color of the rainbow "spiritualizes (*durchgeistigt*) and gives a soul (*beseelt*) to nature." It was not by accident that the great German Renaissance artist Matthias Grünewald painted his angels with the colors of the rainbow, "so that the souls of these holy figures shine as imagination."[2]

In another writing from this same period – "Dialogue on the Religiosity of the Present" (1913) – Benjamin pays homage to Romanticism for its "powerful insights into the nocturnal side of nature." The friend who dialogues with the author in this essay refers to "neo-romanticism," and mentions "appealing but dangerous writers" such as Schnitzler, Hoffmannsthal, and Thomas Mann. But for the author (the "Ich" in the dialogue) the crucial issues of our epoch are to be found elsewhere – in the possibility of a new religion and a new socialism, whose prophets are Tolstoy, Nietzsche, and Strindberg.

This "social religion" is radically opposed to the current, present-day conception of the social, which reduces it to "a mere thing of civilisation, such as the electric light." In this context Benjamin expresses several classic *topoi* of Romantic *Zivilisationskritik*: the transformation of human beings into "work-machines," the degradation of labor into mere technology, the hopeless submission of individuals to social mechanics, and above all the replacement of the "heroic-revolutionary efforts" of the past by the pitiful march of progress, the "crab-like walk of evolution."[3]

This last remark is most revealing of the twist that Benjamin gives to the romantic tradition. The deconstruction of the ideology of progress is not made in the name of conservativism or restoration, but of *revolution*. This subversive shift is also the leading motif of his conference "The Life of Students" (1915), an astonishing document that seems to collect in a single beam of light all the ideas that haunted him later in life. For Benjamin, the true questions that should be raised today are not the "problems of a limited and specialised philosophical science," but rather the "metaphysical questions of Plato, Spinoza, the Romantics, and Nietzsche." Among these "metaphysical" problems, historical temporality is prominent. Against the "formless progressive tendency" with its linear conception of time, Benjamin praises the critical power of *utopian images*, like the French revolutionary idea of 1789 and the messianic Kingdom. Another sort of utopian moment is to be found in the "Tolstoyan spirit" of service to the poor, which has grown "in the ideas of the most profound anarchists and in the Christian monastic communities." In a typically romantic/revolutionary shortcut, the religious past and the utopian future are associated, under the common inspiration of the Russian, libertarian socialist and Christian writer.[4]

Several other of Benjamin's early writings bear witness to his interest in nineteenth century German romanticism: an essay on Hölderlin (1914–1915) and of course his doctoral thesis, *The Concept of Art Criticism in German Romanticism* (1919). But as we have seen, his attraction for the subject was much more than academic. As he explained in a letter of June 1917 to Gershom Scholem, he not only loved the "infinite deepness and beauty" of the early romantics but also considered romanticism as without any doubt the last movement that "saved tradition for us."[5]

Benjamin's political views in those first years were close to anarchism. A whole group of anarchist/romantic writers formed a sort of *constellation* with Benjamin's social philosophy in this period – a group that included not only figures such as Tolstoy and Strindberg (quoted in a letter in 1915), but also Charles Péguy (Benjamin wrote in a letter to Scholem on September 15, 1919, that no writer had ever touched him so closely and fraternally), Georges Sorel (highly praised in an essay on violence), and Gustav Landauer. It was only later that Benjamin discovered Marxism and the communist movement. His first reference to communism appears in 1921, in his Sorelian essay *Critique of Violence*, where he celebrates the "devastating and on the whole justified" critique of parliaments by the Bolsheviks and the anarcho-syndicalists.[6] This link between communism and anarchism was to be an important aspect of his political evolution, since his Marxism to a large extent took on a libertarian coloring.

But it was after 1924, when he read Lukács's *History and Class Consciousness* (1923) and discovered practical communism through the eyes of Asja Lacis – a Soviet artist and political activist he met (and fell in love with) in Capri – that Marxism became a key component of his worldview. In 1929 Benjamin still refers to Lukács's *opus* as one of the few books that remain lively and topical, seeing it as:

> the most fully achieved philosophical work of the Marxist literature. Its uniqueness lies in the assurance with which it grasps the critical situation of class-struggle in the critical situation of philosophy, and in the coming concrete revolution the absolute presupposition, even the absolute implementation and final word of theoretical knowledge. The polemic against it by the hierarchy of the Communist Party under the leadership of Deborin confirms in its way the import of the book.[7]

This commentary illustrates Benjamin's independence of mind in relation to the "official" doctrine of Soviet Marxism, in spite of his sympathy for the USSR.

The first work in which the influence of Marxism can be felt is *One-way Street*, written between 1923 and 1925, published in 1928. Benjamin's former neo-romantic criticism of progress is now charged with a revolutionary Marxist tension, as in the section of the book entitled "Fire Alarm":

> [I]f the abolition of the bourgeoisie is not completed before an almost calculable moment in economic and technical development (a moment signaled by inflation and poison-gas warfare), all is lost. Before the spark reaches the dynamite, the lighted fuse must be cut.

Will the proletariat be able to fulfill this historical task? The survival or destruction of "three thousand years of cultural development" depends on the answer.[8] Benjamin was wrong about inflation, but not about the war, although he could not foresee that the lethal gas would be used not on the battlefields, as in the First World War, but in the industrial extermination of Jews and Gypsies.

Another important argument in *One-way Street* (in the fragment "To the Planetarium") involves sharp criticism of the modern idea of the domination of nature. While human communities of ancient times had an authentic relationship – rooted in an experience of "ecstatic trance" – with the cosmos, the capitalist ruling class uses technology to wage war, opening trenches as "sacrificial shafts" in the womb of "Mother Earth." Under capitalism the only possible relation to nature is domination: "The mastery of nature, so the imperialists teach, is the purpose of all technology." Against this disastrous conception Benjamin proposed a new understanding of technology as "the mastery of relations between nature and humanity."[9] Similarly, in another passage (the section entitled "Imperial Panorama") of this first Marxist essay, Benjamin again emphasizes the contrast between ancient and modern attitudes towards "Mother Earth":

> The earliest customs of peoples seem to send to us a warning that in accepting what we receive so abundantly from nature we should guard against a gesture of avarice. For we are able to make Mother Earth no gift of our own. It is therefore fitting to show respect in taking, by returning a part of all we receive before laying hands on our share.[10]

Benjamin mentions, as a survival of this ancient tradition, the immemorial prohibition against gathering forgotten ears of corn, which then return to the earth. It is curious that Marx's concept of the "metabolic rift," developed in *Capital*, is based on a very similar argument:

> Capitalist production ... disturbs the metabolic interaction (*Stoffwechsel*) between man and the earth, i.e. prevents the return to the soil of its constituent elements consumed by man in the form of food and clothing; hence it hinders the operation of the eternal natural conditions for the lasting fertility of the soil.[11]

Benjamin concludes this passage with a warning that has become even more relevant today than in his own time: "If society has so degenerated through necessity and greed that it can now receive the gifts of nature only rapaciously, … the earth will be impoverished and the land yield bad harvests."[12]

As we saw, Benjamin, in opposition to the vulgar evolutionist brand of Marxism, does not conceive the proletarian revolution as the "inevitable" result of economic and technical progress but as the critical interruption of an evolution leading to catastrophe. This critical standpoint explains why his Marxism has a peculiarly pessimistic spirit – a revolutionary pessimism that has nothing to do with resigned fatalism. In his article on Surrealism in 1929 (*Surrealism: The Last Snapshot of the European Intelligentsia)*, in which he again tries to reconcile anarchism and Marxism, Benjamin defines communism as the organization of pessimism, which means "mistrust in the fate of freedom, mistrust in the fate of European humanity." He then adds this ironic conclusion: one can have "unlimited confidence only in IG Farben and the peaceful perfecting of the Luftwaffe."[13]

Benjamin's critical vision allowed him to perceive – intuitively, but with surprising acuity – the catastrophes in store for Europe as a result of the crisis of industrial/capitalist civilization. But not even Benjamin, one of the most pessimistic of his time, could have foreseen the destruction that the Luftwaffe was to rain down on the civilian populations of Europe's towns and cities, or still less imagine that IG Farben, the great German chemical corporation, just twelve years later would set up plants in the concentration camps to exploit prisoners as forced labor, and produce, through a subsidiary, the lethal gas used in the Final Solution.

The 1929 article attests to Benjamin's interest in Surrealism, which he sees as a modern manifestation of revolutionary romanticism. We might perhaps define the approach common to Walter Benjamin and André Breton as a kind of "Gothic Marxism," distinct from the dominant version that was metaphysically materialistic in tendency and contaminated by the evolutionary ideology of progress. The adjective "Gothic" should be understood in its romantic sense, as fascination with the marvelous and also with the enchanted aspects of pre-modern societies and cultures. The English Gothic novel of the eighteenth century, and some of the German romantics of the nineteenth, are "Gothic" references one finds at the heart of the work of both thinkers. The Gothic Marxism common to them might be said, then, to be a historical materialism sensitive to the magical dimension of past cultures, to the "dark" moment of revolt, or the lightning flash that illuminates the sky of revolutionary action.[14]

It is therefore clear that Benjamin's adoption of Marxism did not at all mean he had lost interest in the romantic protest against bourgeois civilization or the romantic nostalgia for an idealized past. On the contrary, he integrated both into his *sui-generis* Marxist criticism of the capitalist forms of alienation. For instance, in a piece from 1930 he refers to E.T.A. Hoffmann as a romantic follower of the "oldest cultural heritage of humanity," who sought an active link with the most ancient times *(der fernsten Urzeit)*. As we shall see, this reference to a primeval, archaic, or ancestral era was to become central in Benjamin's later writings, in contrast to the common romantic nostalgia for the Middle Ages. Moreover, Benjamin was fascinated by the "most decidedly religious dualism" between "life" and the "automaton" that can be found in the fantastic tales of E.T.A. Hoffmann, Edgar Alan Poe, Alfred Kubin, and Oskar Panizza. According to Benjamin, Hoffmann's *Erzählungen* – he is surely thinking of the famous tale of the mechanical doll Olympia – are inspired by the feeling of a secret identity between the *automat* and the *satanic*, and by the perception of daily human life in modern society as "the product of an infamous artificial mechanism, whose core is regulated by Satan."[15]

This characteristically romantic terror of the mechanization of life is reformulated in contemporary terms by Benjamin in his essays from the 1930s on Baudelaire, in reference to the transformation of proletarians into automatons: the repetitive, senseless mechanical gestures of the worker submitted to the machine (and here Benjamin directly quotes several passages from *Capital*) are similar to the automaton-like gestures of people walking in city streets, as described by Poe and Hoffmann. Both workers and members of the crowd are victims of urban and industrial civilization, which has destroyed all authentically human *Erfahrung* (experience) – rooted in the memory of a cultural and historical tradition – to replace it by a shallow *Erlebnis* (immediate experience), in particular by *Chockerlebnis* (shock experience), which provokes among individuals a purely reactive behavior, transforming them into automatons that have completely lost all memory.[16]

Marxism against the murder of nature (1935–1939)

In Benjamin's writings from the 1920s there are very few references to Marx himself (or Engels). Benjamin does not seem to have had any real knowledge of Marx's writings in that period, and his appropriation of historical materialism is then mainly based on contemporary Marxist literature, not on the texts of the founding fathers. The effective study of these texts appears to have taken place in the 1930s, during his years of exile in Paris as a refugee from Nazi Germany (1933–1940), in connection with his work on the Arcades Project (*Passagen-Werk*). The exact nature of this project remains unclear. Was it to be a new form of book, composed as a montage, a huge assembly of quotations peppered with comments? Or was the collection of files just the raw material, to be used for writing a book that never came into being? In any case, it documents Benjamin's intensive study of Marx's and Engels's writings after 1934, as well as his highly selective and idiosyncratic approach.

For a brief "experimental" period between 1933 and 1935, during the years of the Second Five-Year Plan, some of Benjamin's Marxist texts seem close to "Soviet productivism" and an uncritical adherence to the promises of technological progress. The main texts of this kind are "Experience and Poverty" (1933), "The Author as Producer" (1934), and, though only to a degree, "The Work of Art in the Age of Mechanical Reproduction" (1935). However, even in those years Benjamin had not quite lost his interest in the romantic problematics, as his 1935 article on Bachofen attests. In fact, Benjamin's thinking in this period is quite contradictory. He sometimes shifts very rapidly from one extreme to the other – even in a single text, as in the famous essay on the work of art. One finds in these writings both a permanent aspect of his Marxist thinking – the materialist dynamic – and an "experimental" tendency to push certain arguments to their ultimate consequences. He seems tempted by a Soviet variant of the ideology of progress, though reinterpreted in his own way. Some Marxist readings of Benjamin's works foreground just these texts that seem closer to a "classic" – if not orthodox – historical materialism. After 1936 this kind of "progressive parenthesis" closes and Benjamin increasingly reintegrates the romantic moment into his idiosyncratic Marxist critique of the capitalist forms of alienation. This applies also to the issue of technology's relation to nature. For instance, in the French version of the work of art essay, written in 1936, he distinguishes between two sorts of technology, one aiming at the "enslavement (*asservissement)* of nature" and the other at a "harmony between nature and humanity" – the concept of *harmonie* being, of course, borrowed from Fourier. In the essay on *The Storyteller* Benjamin gives a typical romantic example of the "complicity" between humans and nature: the fairy tale. But this *Komplizität* requires human emancipation: "The liberating magic which the fairy tale has at its disposal does not bring nature into play in a mythical way, but points to its complicity with liberated man."[17]

The *Passagen-Werk* belongs, in most of its sections, to this heterodox "romantic Marxist" project. The aim of the Arcades Project is defined by Benjamin in the following terms:

> One can see as one of the methodological aims of this work to demonstrate the possibility of a historical materialism that has annihilated in itself the idea of progress. Here is precisely where historical materialism has to dissociate itself from the bourgeois habits of thought.[18]

Such a program did not aim at some sort of "revision" but rather, as Karl Korsch tried to do in a book that Benjamin comments on, a return to Marx himself.

One of the aspects of this "annihilation" is a new interpretation of Marx's intellectual sources, emphasizing his relation to the romantic critique of civilization. Benjamin mentions Korsch approvingly in this respect:

> Korsch says very correctly – and one might think here of De Maistre and De Bonald: "Therefore, into the theory of the modern labor movement also entered … a part of that … 'disillusion' which followed the great French Revolution, immediately proclaimed by the first French theoreticians of the counter-revolution, and then by the German Romantics, which exerted a powerful influence, especially though Hegel, on Marx."[19]

One may doubt that Joseph De Maistre – who is extensively quoted in the Arcades section on Baudelaire – was of any interest to Marx, who probably never read him. But the general hypothesis that romantic anti-bourgeois currents were relevant for Marx is quite appropriate, and of course corresponds to Benjamin's own attempt to reformulate historical materialism.

These romantic undercurrents are also mentioned in another extensive quote from Korsch's book:

> Sources of Marx and Engels: "From the bourgeois historians of the French Restoration they took the historical importance of class and class struggle; from Ricardo the conflicting economic interests of the social classes; from Proudhon, the description of the modern proletariat as the only revolutionary class; from the feudal and Christian assailants of ... the new economic order, the ruthless unmasking of the liberal ideas of the bourgeoisie, the piercing invective full of hatred. Their ingenious dissection of the modern mode of production they took from the petty-bourgeois socialism of Sismondi; the accents of humanism perceptible even in their later materialistic writings, from earlier companions among the left Hegelians, especially Feuerbach; the relevance of politics to the struggle of the working class, from the contemporary labor parties, French Social-Democrats and English Chartists; the doctrine of revolutionary dictatorship from the French Convention and from Blanqui and his followers. Finally they took from Saint Simon, Fourier and Owen the ultimate goal of all socialism and communism, the complete overthrow of existing capitalist society, abolition of all classes and class oppositions, and transformation of the political State into a mere management of production."[20]

For both Korsch and Benjamin, romantic critics – from "Christian socialists" to Sismondi – occupy a significant place in this genealogy of Marxist theory.

Another argument in Benjamin's attempt to emancipate Marxism from the illusions of progress is a critique of the idealization of industrial labor. Several quotations from Marx or Engels in the Arcades project are linked to this criticism, for instance the one in which Engels, in *The Condition of the Working Class in England* (1845), compares the "infinite torture of labor, where the same mechanical process is again and again repeated," with the infernal labors of Sisyphus: "the weight of labor, like the rock, falls again and again on the backs of the exhausted worker."[21] In a similar vein, once more using the infernal allegory, Benjamin refers to Marx's well-known comparison of the capitalist factory with Dante's Inferno section of the *Divine Comedy*, in which the gate to Hell has at its entrance the sinister phrase *Voi che entrate, lasciate qui ogni speranza* (Abandon all hope, ye who enter here). He is also persuaded that the decline of human experience (*Erfahrung*) started with the first modern factories (*Manufaktur*) at the beginning of commodity production. Benjamin's argument, here as elsewhere in his work, is a revised version of the Romantic aversion for modern machinism, industrial labor and capitalist technology, but formulated in Marxist terms. It is a polemic against bourgeois discourse, but also, implicitly, against the leftist versions of the modern cult of industry and labor, and more generally against the ideology of "progress" so deeply ingrained in the socialist culture of his time.[22]

Benjamin's Marxism, as developed in the Arcades project and in his last writings, constitutes a new and original reinterpretation of historical materialism, one radically different from the orthodoxy of the Second and Third Internationals. It should be seen as an attempt to deepen and radicalize the opposition between Marxism and bourgeois ideology, to heighten its revolutionary potential and sharpen its critical content. Politically Benjamin was, during the second half of the 1920s and the 1930s, an idiosyncratic sympathizer with the communist movement. However, he felt a strong affinity with Leon Trotsky – as attested by a letter he wrote to Gretel Adorno in the Spring of 1932 – and, particularly after 1937, increasingly distanced himself from Soviet (Stalinist) Marxism. From this *linksradikal* (left-radical) commitment followed quite logically a very critical assessment of social democracy, the blindness of which he confronted with the powerful insights of Marx and Engels. For instance, his article of 1937 on "Edward Fuchs, Collector and Historian" contains a sharp attack on the social-democratic ideology, which combines Marxism with positivism, Darwinist evolutionism, and the cult of "progress." In his view its greatest mistake is that it sees in the development of technology only the progress of the natural sciences, and not the social regression. It never perceives the danger that the energies produced by technology can serve above all for the technical perfection of war. Against the shallow optimism of the social-democratic pseudo-Marxists, Benjamin opposes his pessimistic-revolutionary perspective, citing "the perspective on early barbarism that illuminates Engels's *The Condition of the Working Class in England* and Marx's own prognosis on the development of capitalism."[23]

Although Benjamin rejected the doctrines of progress, this did not prevent him from positing a radical alternative to the impending disaster – a revolutionary utopia. In *Paris, Capital of the Nineteenth Century* (1935) he argues that utopias – dreams of a different future – are closely associated with elements coming from prehistory (*Urgeschichte*), that is, from a primitive, classless society. Stored in the collective unconscious, these experiences of the past "interact with the new to give birth to ... utopias."[24] The romantic protest against capitalist modernity is always, as we have argued, formulated in the name of an idealized past – real or mythical. While in Benjamin's early theological writings there are often references to a lost paradise, in the 1930s it is specifically primitive communism that comes to play this role – as in Marx and Engels, who borrowed the idea from the romantic anthropology of Maurer, Morgan, and Bachofen.

Benjamin's review of Bachofen, written in 1935, is one of the most important keys to understanding his method of weaving a new philosophy of history with the threads of Marxism and romanticism. Rejecting the conservative (Klages) and fascist (Bäumler) interpretations of Bachofen, he claims that the latter's work on matriarchy, "inspired by romantic sources," has attracted the interest of both Marxist and anarchist thinkers because of his "evocation of a communist society at the dawn of history." The anarchist geographer Elisée Réclus found in Bachofen's books the ancient sources of his libertarian ideal, while Engels and Paul Lafargue were interested in his presentation of matriarchal communities as social organizations in which there existed a high degree of democracy and civil equality, as well as forms of primitive communism that subverted the concept of authority.[25]

Archaic societies also lived in greater harmony with nature. In the *Passagen-Werk* Benjamin once more calls into question the "mastery" (*Beherrschung*) of nature and its "exploitation" (*Ausbeutung*) by modern societies. As Bachofen had already shown, Benjamin insists, "the murderous (*mörderisch*) idea of the exploitation of nature" – a dominant capitalist/modern concept from the nineteenth century on – did not exist in matriarchal societies of the past because nature was perceived as a generous mother (*schenkenden Mutter*).[26] It should be emphasized that the German word *Ausbeutung* has somewhat more radical connotations than "exploitation": sharing a common root with *Beute*, "booty," it suggests looting. Thus, *Ausbeutung der Natur* means not only the exploitation but also the plundering of nature, which is of course one of the main critical *topoi* of modern ecology.

These passages from the Arcades Project are perhaps the most important proto-ecological arguments in Benjamin's writings. There is in fact a striking similarity between his views and those expressed at the World People's Conference on Climate Change and the Rights of Mother Earth in Cochabamba, Bolivia (April 2010), convened by President Evo Morales. The resolutions adopted at Cochabamba take up, almost literally, Benjamin's argument about industrial-capitalist civilization's "murderous" relationship with nature, which traditional communities regarded as a "generous Mother."[27] For Benjamin – as for Engels and the libertarian socialist Élisée Reclus – it was a question not of a return to the prehistoric past but of putting forward the prospect of a new harmony between society and the natural environment. The name that sums up for Benjamin the promise of such a future reconciliation with nature is Fourier. Only in a socialist society in which production would no longer be based on the exploitation of human labor, would work:

> no longer be characterized as the exploitation of nature by man. It would be conducted on the model of children's play, which in Fourier forms the basis of the "impassioned work" of the Harmonians … Such work invested with the spirit of play is oriented not towards the production of values but towards the improvement of nature.[28]

Similar ideas are suggested in Benjamin's essays on Baudelaire. For Benjamin the expression "*vie antérieure*" refers to a primitive age (analogous to the one studied by Bachofen), where authentic experience still existed and where ritual ceremonies and holidays allowed for a fusion between the individual and the collective past. It is this memory of an ancient *Erfahrung* that inspires the game of *correspondances* between humanity and nature in Baudelaire, and his refusal of the *modern catastrophe*: "The correspondances are dates of rememoration. They are not historical but pre-historical data. What constituted the greatness and significance of the ancient festivals was the encounter with an earlier life."[29] Clearly Benjamin is not proposing a return to pre-historical times, but rather dreaming of a utopian society the likes of which he seems to glimpse in Paul Scheebart "futuristic" novels – as he suggests in a short French note, "Sur Scheerbart," in 1938 – where "human beings will give up the vile and crass opinion that they are called to 'exploit' the forces of nature," and will instead understand the need for a new form of technology that will liberate humans as well as "fraternally liberate the whole creation."[30]

The revolution is the emergency brake

In 1939, as the war began, Benjamin was interned as an "enemy alien" by the French government. He managed to escape from the internment camp, but after the German victory and occupation of France in 1940 he was obliged to flee Paris for Marseille. In these dramatic circumstances he wrote his last piece, the "Theses on the Philosophy of History," perhaps the most important document in revolutionary theory since Marx's celebrated *Theses on Feuerbach* (1845). A few months later, in September 1940, after a failed attempt to escape through Spain, he chose suicide.

In these few but extraordinarily dense pages Marx is often quoted once more as the thinker of class struggle and of revolution. The ideology of progress – as it was to be found inside the communist movement as well as in bourgeois society – is there criticized in its philosophical foundations, in the form of the concept of linear and empty time as opposed to Benjamin's own messianic conception. The question of the relationship between Marxism and messianism in Benjamin's late writings is of course a most controversial one. During the sharp polemics of the 1960s in Germany some insisted on the religious dimension, others on his Marxist materialism. Benjamin himself referred ironically (in a letter to Scholem) to his "Janus face," but the critics often looked at one face only, ignoring the other. In order to go beyond this kind of polemic, it is useful to recall that the Roman god had two faces but only one head. Benjamin's faces are manifestations of one single thought process, which had simultaneously a messianic and a Marxist expression.

As in previous writings mentioned above, here too Benjamin forcibly argues – very much against the current in relation to the dominant views of the Left in his time – that the suppression of labor exploitation (the traditional aim of the socialist labor movement) cannot take place if nature continues to be exploited. Interestingly, Marx had already suggested in *Capital* that the two exploitations are linked: "Capitalist production, therefore, only develops the technique and the degree of combination of the social process of production by simultaneously undermining the original sources of all wealth – the soil and the worker."[31] Benjamin was probably not familiar with this passage, which he does not mention.

This issue is mainly discussed in Thesis XI, where Benjamin emphasizes the radical opposition between Marx and (German) Social Democracy concerning the treatment of nature. For instance, denouncing the Social-Democratic idealization of industrial labor – a criticism already made in the Arcades Project – Benjamin comments:

> Nothing has so corrupted the German working class as the notion that it was moving with the current. It regarded technological development as the driving force of the stream with which it thought it was moving. From there it was but a step to the illusion that the factory work ostensibly furthering technological progress constituted a political achievement. The old Protestant work ethic was resurrected among German workers in secularized form. The Gotha Program already bears traces of this confusion, defining labor as "the source of all wealth and all culture." Smelling a rat, Marx countered that "the man who possesses no other property than his labour power" must of necessity become "the slave of other men who have made themselves owners."[32]

As an attentive reader of Max Weber, Benjamin believed that the Protestant work ethic had close connections – by elective affinity – with the spirit of capitalism. Benjamin draws on both Weber and Marx to criticize the conformist posture of Social Democracy in relation to industrial-capitalist production.

He also denounces what he considers the "corrupt" Social-Democratic concept of nature, such as it appears in the comment by Joseph Dietzgen – a now-forgotten nineteenth-century "socialist philosopher" who was quite influential in German Social Democracy – that nature is "here gratis," in other words a cheap commodity that can be plundered indefinitely. For Benjamin – in a comment not included in the "Theses" – Dietzgen's idea of a "gratis disposable nature" is synonymous with the sinister idea of "exploiting nature."[33] In the same Thesis XI Benjamin once again hails Fourier as the utopian visionary of "a labor that, far from exploiting nature, is capable of extracting from it the virtual creations that lie dormant in her womb." The examples he gives for Fourier's utopian dream are curious. Some are poetic – four moons would illuminate the night, sea water would not taste salted – but others seem, from a twenty-first century perspective, rather disastrous: "the ice will retreat from the Poles" (exactly what is happening today, of course, with global warming). But instead of reading them literally one should perhaps perceive them as expressions of the desire to transform the planet into an *enchanted garden*.

It would be a mistake to suppose that Benjamin wanted to replace Marxism with utopian socialism. He regarded Fourier as an (important) complement to Marx. Both were opposed to Social-Democratic positivism – again typified by Joseph Dietzgen – for whom "the new conception of labor amounts to the exploitation of nature, which with naive complacency is contrasted with the exploitation of the proletariat." This is "a conception of nature which differs ominously from that of the Socialist utopias before the 1848 revolution," and one which "already displays the technocratic features later encountered in Fascism."[34] This last comment is particularly interesting, because German fascism pretended, in its propaganda, to resurrect a traditionalist view of nature and of the natural organic bonds of "Blood and Soil" (*Blut und Boden*). Benjamin, instead, emphasizes the *reactionary modernism* of the technocratic industrial (capitalist) project of the Third Reich, based on ruthless exploitation of natural resources, which was hidden behind this "organic" façade.

As we saw in the Arcades Project, the socialist utopia proposed by Marx and Fourier has its origins in the ancient past – the prehistoric Garden of Eden of primitive communism discussed by Bachofen, Engels, and other socialists. How does it relate to the Lost Paradise of the famous Thesis IX of "Theses on the Philosophy of History"? One should interpret this enigmatic and fascinating text as an allegory in which each sacred image has a profane "correspondent" (in the Baudelairian sense). History is represented as a powerless angel, inexorably projected into the future by a storm that blows from paradise, while at his feet ruins and wreckage accumulate: "What we call progress is *this* storm."[35] It is difficult to avoid the conclusion that the paradise from which we are being blown away by the catastrophe of progress has its profane equivalent – or rather its "correspondent" – in the egalitarian prehistoric society, the primitive community that was dreamt not only by the historian of matriarchy (Bachofen), but also the *poète maudit* (Baudelaire) and the founding fathers of socialism.

As we can see, the "recasting" of historical materialism in the *Theses* involves a selective – and heterodox – reappropriation of the Marxian themes that seem to Benjamin essential to his undertaking: the state as class domination, the class struggle, the social revolution, and the utopia of a classless society. Materialism itself, revised by theology, is incorporated into his theoretical system, as well as key elements of the romantic *Zivilisationskritik*. The result is a reworking, a critical reformulation of Marxism, that integrates messianic, romantic, Blanquist, anarchist, and Fourierist "splinters" into the body of historical materialism. Or rather it is the fabrication, using all these materials, of a new and heretical Marxism, radically different from all the – orthodox or dissident – variants of Benjamin's time.

There are hardly any direct criticisms of Marx and Engels in the *Theses* themselves, but they do figure here and there in the associated notes. At one important point Benjamin marks a critical distance from the author of *Capital* on the issue of "progress":

> Critique of the theory of progress in Marx. Progress is here defined as the development of productive forces. But the human being, e.g. the

> proletariat, belongs to them. In this way the question of the criteria is only pushed back.[36]

This is actually a point of major significance, since an uncritical view of the development of productive forces largely fueled the economistic interpretations of the Second International and of Stalinist productivism, in contrast to an ecological approach. But the issue remains at the level of a programatic proposal, and Benjamin does not go into it more deeply.

Another critical comment is essential to understanding Benjamin's conception of revolution: "Marx said that revolutions are the locomotive of world history. But perhaps things are very different. It may be that revolutions are the act by which the human race traveling in the train applies the emergency brake."[37] Implicitly, the image suggests that if humanity allows the train to follow the course already laid down by the steel structure of the rails, and nothing stops its vertiginous career, we shall be hurled into catastrophe – a crash into the abyss. This passage is one of the preparatory notes for "Theses on the Philosophy of History," and does not appear in the final versions of the document. The passage from Marx to which Benjamin refers appears in *The Civil War in France*: "Die Revolutionen sind die Lokomotiven der Geschichte" (the word "world" does not appear in Marx's text).

Of course, even Benjamin, the most pessimistic of Marxists, could not foresee how far the process of capitalist exploitation and domination of nature – and its bureaucratic double in the countries of Eastern Europe before the fall of the Wall – would have catastrophic consequences for all humanity. We are witnessing now in the early twenty-first century the ever faster "progress" of the train of industrial/capitalist civilization into the abyss, an abyss called ecological disaster, the most dramatic expression of which is global warming. It is important to bear in mind the increasing acceleration of the train, the dizzying speed at which it is racing towards catastrophe. Some years ago, when one referred to the dangers of ecological catastrophe, it was in the distant future, perhaps at the end of the twenty-first century. Hans Jonas, in *The Imperative of Responsibility* (1984), called on us to protect the lives of generations yet unborn. Now, though, the process of climate change has accelerated to the point that we are discussing what will happen in the next few decades. In fact, the catastrophe has already begun and we are in a race against time to try to impede, slow down, and contain the process, which will result not only in the raising of the planet's temperature but the desertification of vast areas, the rising of sea levels, and the disappearance under the waves of coastal cities: Venice, Amsterdam, Hong Kong, and Rio de Janeiro.

A revolution is necessary, Benjamin wrote, to halt the race towards catastrophe. Ban Ki-moon, then Secretary-General of the United Nations, who is far from being a revolutionary, offered the following diagnosis (*Le Monde*, September 5, 2009): "We," he said, referring no doubt to the governments of the world, "have our foot stuck on the accelerator and we are heading towards an abyss." Benjamin defined the destructive progress that accumulates catastrophes as a "storm." The same word "storm" appears in the title (which seems to have been inspired by Benjamin) of the latest book by James Hansen, a NASA climatologist and one of the world's foremost specialists on climate change. Published in 2009, the title of the book is *Storms of my Grandchildren. The Truth about the Coming Climate Catastrophe and Our Last Chance to Save Humanity*. Hansen is no revolutionary either, but his analysis of the coming "storm" – which is for him, as for Benjamin, an allegory for something much more menacing – is impressive in its lucidity.

Will humanity apply the revolutionary brakes? Every generation, Benjamin writes in "Theses on the Philosophy of History," has been endowed with a "weak Messianic power," and ours is no exception. As he had written already in *One-way Street* in the 1920s, if we do not exercise it "before an almost predictable moment in economic and technological development, ... all is lost." Like Thomas Cole a century earlier, Benjamin was a prophet, not as a diviner of the future, or Greek oracle, but in the Old Testament sense of one who calls the people's attention to future dangers. His predictions were conditional: see what will happen, unless ... if we do not ... In this perspective, the future is still open. In a metaphor from Jewish tradition that Benjamin made his own, every second is the narrow gate through which salvation can come.

Notes

1 Benjamin, *Gesammelte Schriften*, II, 1, 46. All translations from Benjamin's works are ours, unless otherwise indicated.
2 Benjamin, *Nachträge*, in *Gesammelten Schriften*, VII, 1, 24–25.
3 Benjamin, *Gesammelte Schriften*, II, 1, 19–20, 25, 34.
4 Benjamin, *Gesammelte Schriften*, II, 1, 75, 79, 82.
5 Benjamin, *Briefe*, I, 138.
6 Benjamin, *Gesammelte Schriften*, II, 1, 191.
7 Benjamin, *Gesammelte Schriften*, III, 171.
8 Benjamin. *Reflections: Aphorisms, Essays and Autobiographical Writings*, 84.
9 Benjamin, *One-way Street and Other Writings*, 87.
10 Benjamin, *One-way Street and Other Writings*, 60.
11 Marx, *Capital: Volume One*, 637–638.
12 Benjamin, *One-way Street and Other Writings*, 60.
13 Benjamin, *Gesammelte Schriften*, II, 1, 308.
14 See Cohen, *Profane Illumination: Walter Benjamin and the Paris of Surrealist Revolution*, 1–2; and Löwy, *Morning Star: Surrealism, Marxism, Anarchism, Situationism, Utopia*.
15 Benjamin, *Gesammelte Schriften*, II, 2, 642–643, 646–647.
16 Benjamin, *Passagen-Werk*, in *Gesammelte Schriften*, V, 2, 966; and Benjamin, *Gesammelte Schriften*, I, 2, 632–636.
17 Benjamin, *The Storyteller*, 11.
18 Benjamin, *Passagen-Werk*, in *Gesammelte Schriften*, V, 1, 574.
19 Benjamin, *Passagen-Werk*, in *Gesammelte Schriften*, V, 2, 820.
20 Karl Korsch, *Karl Marx* (1938), as quoted in Benjamin, *Passagen-Werk*, in *Gesammelte Schriften*, V, 1, 819–820.
21 Benjamin, *Passagen-Werk*, in *Gesammelte Schriften*, V, 1, 162.
22 Benjamin, *Passagen-Werk*, in *Gesammelte Schriften*, II, 813, 966.

23 Benjamin, *Gesammelte Schriften*, II, 2, 488.
24 Benjamin, *Gesammelte Schriften*, V, 1, 47.
25 Benjamin, *Gesammelte Schriften*, II, 1, 220–230.
26 Benjamin, *Passagen-Werk*, in *Gesammelte Schriften* V, 1, 456.
27 For the final declaration of the Cochabamba conference, the "People's Agreement," see World People's Conference on Climate Change and the Rights of Mother Earth (website). Accessed April 9, 2019. https://pwccc.wordpress.com/2010/04/24/peoples-agreement.
28 Benjamin, *Passagen-Werk*, in *Gesammelte Schriften* V, 1, 456.
29 Benjamin, *Gesammelte Schriften*, I, 2, 639.
30 Benjamin, *Gesammelte Schriften*, II, 2, 631.
31 Marx, *Capital: Volume One*, 637–38.
32 Banjamin, *Illuminations*, 258–259.
33 Benjamin, *Über den Begriff der Geschichte*, 22, 144.
34 Benjamin, *Über den Begriff der Geschichte*, 22.
35 Benjamin, *Über den Begriff der Geschichte*, 35–36.
36 Benjamin, *Über den Begriff der Geschichte*, 130.
37 Benjamin, *Über den Begriff der Geschichte*, 153.

Works Cited

Benjamin, Walter. *The Arcades Project*. Translated by Howard Eiland. Edited by Kevin McLaughlin. Cambridge, MA: Harvard University Press, 1999.

Benjamin, Walter. *Briefe*. Edited by Gershom Scholem and Theodor W. Adorno. Frankfurt am Main: Suhrkamp Verlag, 1966.

Benjamin, Walter. *Gesammelte Schriften*. Edited by Rolf Tiedemann and Hermann Schweppenhäuser. 7 Vols. Frankfurt am Main: Suhrkamp Verlag, 1972–89.

Benjamin, Walter. *Illuminations*. Translated by H. Zohn. New York: Schocken, 1969.

Benjamin, Walter. *One-way Street and Other Writings*. Translated by J.A. Underwood. London: Penguin, 2008.

Benjamin, Walter. *Reflections: Aphorisms, Essays and Autobiographical Writings*. Translated by E. Jephcott. New York: Harcourt, Brace, 1978.

Benjamin, Walter. *The Storyteller: Tales out of Loneliness*. Translated by Sam Dolbear, Esther Leslie and Sebastian Truskolaski. London: Verso, 2016.

Benjamin, Walter. *Über den Begriff der Geschichte*. Edited by Gerard Raulet. Frankfurt am Main: Suhrkamp, 2010.

Cohen, Margaret. *Profane Illumination: Walter Benjamin and the Paris of Surrealist Revolution*. Berkeley, CA: University of California Press, 1993.

Hansen, James. *Storms of my Grandchildren: The Truth about the Coming Climate Catastrophe and Our Last Chance to Save Humanity*. New York: Bloomsbury, 2009.

Jonas, Hans. *The Imperative of Responsibility*. Chicago, IL: University of Chicago Press, 1984.

Löwy, Michael. *Morning Star: Surrealism, Marxism, Anarchism, Situationism, Utopia*. Austin, TX: University of Texas Press, 2009.

Marx, Karl. *Capital: Volume One*. Translated by Ben Fowkes. New York: Vintage, 1977.

5 Raymond Williams: romantic culture and socialist ecology

Son of a Welsh railway worker, Raymond Williams (1921–1988) was always something of an outsider in the British academic elite, by virtue of his double loyalty, throughout his entire life, to the working class and to Welsh culture. To be sure, he studied literature at the very elitist Cambridge University, but by joining the Communist Party at the end of the 1930s he placed himself necessarily at the margins of the academic establishment. His studies were interrupted by the Second World War, in which he fought in an armored tank unit. After the war he decided to leave Cambridge – as well as the Communist Party – and began to lecture, with his friend E.P. Thompson, at the Workers' Educational Association, a pedagogical network linked with the British Labour movement. He only returned to Cambridge, as a lecturer, in 1961.

Williams, together with Stuart Hall and their friends at the Birmingham Centre for Contemporary Cultural Studies, was one of the primary initiators – through the influence of his book *Culture and Society* (1958) – of "Cultural Studies," which was destined to become one of the most important areas of Anglo-American academic research. Williams was also, along with E.P. Thompson, John Saville, Stuart Hall, Raphael Samuel, and other British Marxist intellectuals who left the Communist Party in 1956, a founder of *New Left Review*. A number of his works – among others *Keywords: A Vocabulary of Culture and Society* (1976) and *Marxism and Literature* (1977) – nourished the reflection of several generations of critical intellectuals on both sides of the Atlantic.

His influence was considerable. It is enough to mention that at the end of the 1970s 750,000 copies of his books had been sold. His first writings dealt exclusively with British culture, but during the 1960s he discovered Western Marxism – the historicist/humanist tendency that includes Gramsci, Lukács and Lucien Goldmann – with which he felt strong affinities. His work became a bridge between the British romantic cultural tradition and Continental Marxism. A committed intellectual with strong socialist beliefs, he published, with E. P. Thompson, an influential anti-capitalist document, the *May Day Manifesto* (1968), which categorically denounced the myths of "modernization":

> As a model of social change, modernization crudely foreshortens the historical development of society ... The whole past belongs to

> "traditional" society, and modernization is a technical means for breaking with the past without creating a future ... It is a technocratic model of society, conflict-free and politically neutral, dissolving genuine social conflicts and issues in the abstractions of "the scientific revolution," "consensus," "productivity."[1]

The author of novels, works of sociological analysis, Marxist studies of literature, theater, and media (especially television), as well as political essays, Williams was also one of the main proponents of the Campaign for Nuclear Disarmament (CND). It is important to emphasize that he, unlike many former leftists, never gave up his radical ideas and socialist commitment. This defining characteristic of his work was obfuscated in many later Anglo-American "cultural studies," which often attempted to interpret his writings according to a "post-modern" and "post-Marxist" approach that eliminated the critical, anti-capitalist thrust of his thought.

The romantic dimension is quite crucial for understanding Raymond Williams's work, which spans a very long period – from the late 1940s to the late 1980s. Williams undeniably evolved over the course of his career, but he never repudiated the romantic underpinnings of his thought. *Marxism and Literature* (1977) is doubtless the least romantic of his writings, although it does not contradict his earlier positions, as is corroborated by other texts dating from the late 1970s.[2] In the 1980s Williams reacted more and more strongly against structuralism and post-structuralism, and again brought the revolutionary romantic perspective to the forefront of his thinking, developing it in new ways, including an ecological emphasis.

We will focus first on a few important examples of Williams's non-fictional work in the social and human sciences.[3] Since they are taken from different periods in his career, they will allow us to show both the continuity and the evolution of Williams's romantic problematics.[4]

Culture and Society

In *Politics and Letters* (1979), a series of interviews with several editors of *New Left Review*, one of the interviews deals with *Culture and Society*, the masterpiece which first brought Williams to public attention. Williams explains that the main intention of the work was "oppositional." He wanted:

> to counter the appropriation of a long line of thinking about culture to what were by now decisively reactionary positions, ... to try to recover the true complexity of the tradition it had confiscated – so that the appropriation could be seen for what it was ... The selective version of culture could be historically controverted by the writings of the thinkers who contributed to the formation and discussion of the idea.[5]

The tradition that Williams wished to save from appropriation by the Right (he began the research and reflection that was to culminate in *Culture and Society* in 1948, at the beginning of the Cold War) was precisely the romantic tradition in English letters, one that extends from the late eighteenth century up to the contemporary period.

The study takes as its point of departure the perception that during that time span two parallel developments were taking place: on the one hand, the introduction and evolution of a new meaning of the word "culture," and on the other, the huge social transformations involved in the advent of industrial capitalism. Culture in its new meaning is defined by contrast with the new society; it is the domain of the ideal and of true value, as opposed to the unjust, corrupt, degraded new civilization. Some see culture as a refuge within modern society – the enclave of the intellectual and artistic elite – while for others it is nothing if it is not generalized ("a common culture"). All share the conviction that a cultural community existed before the coming of modernity.

In exploring this tradition, Williams demonstrates the diversity of political positions that are to be found within the same basic posture of refusal of modern commercial, industrial, and mechanistic society. He begins by pointing out the affinities between conservatives such as Burke and Southey, and radicals or socialists like Cobbett and Owen. Later, he locates within the same tradition the anarcho-communist William Morris and T.S. Eliot, who longed to restore medieval Christendom. He also shows that certain authors, like Carlyle and D. H. Lawrence, evolved ideologically while remaining fundamentally romantic. Williams attributes a crucial role to Morris in the development of the critique of "society" in the name of "culture," seeing in him a "pivotal figure" between the nineteenth and twentieth centuries. According to Williams, "in the middle of the twentieth century Morris remains a contemporary thinker, for the directions which he indicated have become part of a general social movement. Yet he belongs, essentially, with the great Victorian rebels ..."[6]

Williams devotes an important chapter of *Culture and Society* to the relationship between the romantic tradition and Marxism (mainly English, for at that time Williams's intellectual horizon was almost exclusively British). He claims that many of the English Marxists of the 1930s were continuing in a different form the tradition of Matthew Arnold and Morris, since their attempt to create a Marxist theory of culture constituted "an interaction between Romanticism and Marx, between the idea of culture which is the major English tradition and Marx's brilliant revaluation of it." When Williams adds, "We have to conclude that the interaction is as yet far from complete," he is, perhaps without knowing it, defining the project that he and E.P. Thompson were to carry out in later years.[7]

While in the main body of the work Williams examines an intellectual and artistic trend in "high culture," analyzing and evaluating it in its own terms, the conclusion introduces a new point of view – that of the working class. Contrary to most of the Oxford Marxists of the 1930s, Williams was born into that class and in the conclusion of his study he brings to bear his personal perspective, attempting to link the romantic tradition in elite culture with the culture of the working class. In both cases he finds an ethos of "community" as opposed to the "bourgeois idea of society" as an aggregate of competing individuals, although the middle-class romantics base this community on an ethic of "service," while for the working class it takes the form of "solidarity."[8] Williams prefers the latter, and his celebration of a working-class ideal here is an example of elements already present in his early work that are more fully developed later.

We can clearly see the continuity in Williams's work also in his response to the friendly but sharp criticisms that the editors of *New Left Review* made in *Politics and Letters* (1979) with regard to *Culture and Society*. In particular, they reproach the book for eliminating the political dimension of the writers studied, and thereby minimizing or masking the objectionable political positions taken by some of them (Carlyle's later apology for racism and imperialism, for instance). In his answer, although Williams acknowledges that he should perhaps have dealt more explicitly with these phenomena, and that he would not write the book in the same way at present, he nonetheless insists on defending his fundamental choices of the 1950s:

> I had discovered themes profoundly related to my sense of the social crisis of my time and the socialist way out of it, not in the approved list of progressive thinkers, but in these paradoxical figures ... [B]y drawing back I was able to reintroduce certain themes and issues which have seemed to me the crucial stuff of action to this day, but which were absent from what I knew then and often know now as politics.[9]

The communists and socialists were often as much the proponents of an exacerbated productivism as were the bourgeois ideologues, and for Williams it was precisely with the post-war explosion of productivity that "the questions posed by Blake or Cobbett acquired their force."[10]

The Country and the City

The same problems concerning modernity are major concerns in *The Country and the City*, written at the beginning of the 1970s at a time when Williams was already coming closer to an explicit Marxist perspective. The book explores a theme in English literature that partially overlaps with that of *Culture and Society* – the contrast between country and city – by analyzing the often complex relationships between the literary images and the real history, notably in terms of class relations and points of view. In *Politics and Letters* the editors of *New Left Review* praise this approach as a step forward in relation to *Culture and Society*. Yet they point out that the book's true importance lies rather in its *transcendence* of the usual Marxist problematics, since in general the work of Marx and Engels "lacks the idea of a *continuing* tradition of values from the past that informs the struggles of the present" (italics in the original text).[11] Williams's distinctive contribution, then, was to have shown the effective presence, through the historical unfolding of the romantic vision, of pre-capitalist values in the modern emancipatory project.

In the interview on *The Country and the City*, Williams calls it a "polemical" book. But interestingly, in different places in the interview he claims to have aimed at two opposite targets. On the one hand, Williams says he was attacking a time-honored interpretation of the poetry celebrating country houses, which saw the poems as simple "records" of life in those houses, "and so of the organic rural society England had once been ...." On the other hand, he declares that his polemic was directed against the dominant tendency in the Labour Party at the time, which conceived socialism as no more than "a successful industrial capitalism without the capitalists."[12] In fact, in *The Country and the City* Williams is battling on two fronts: as a revolutionary, against a form of romanticism that is a reactionary mystification, and as a romantic, against the adulteration of the revolutionary project, its corruption by the modern world.

The book's first moment, then, is a critique of the ideology of the literary tradition praising country life. Williams questions the idyllic image of the country estate projected in seventeenth-century "neo-pastoral" literature, maintaining that in reality those estates were the locus of an early agrarian capitalism. From the mid-eighteenth century on he sees a shift in rural literature from idealization to a melancholic sense of loss. In this later literature he finds a repeated pattern, in which each generation regrets a Golden Age of rural well-being, reliving the illusion of having itself experienced the disappearance of Old England.[13]

In some passages, this demystification of an idealizing or nostalgic thematic of rusticity can give the impression that Williams is carrying on an anti-romantic offensive. This is not the case, however. He is attempting rather to call for a romanticism that can be convincing because it is lucid and complex, capable of engaging the realities of modernity. It might nonetheless seem regrettable that Williams is not more sensitive to the genuinely utopian dimension that could manifest itself even in those expressions of a leisured elite that he is analyzing. But in *The Country and the City* he puts forward his own working-class perspective and experience as he did not in *Culture and Society* (or rather only in the conclusion). He refers to his youth in a rural region of Wales (his father was a worker, but in a small village), and to the tight-knittedness of the local working population. It is the concrete solidarities of this "knowable community"[14] that he affirms as a true rural ideal.

In *The Country and the City* Williams also shows that he is sensitive to the alienation spawned by the modern city. He cites as compelling testimony texts by Wordsworth, Carlyle, and Hardy, as well as Engels's *The Condition of the Working Class in England*, that describe the experience of social fragmentation in the big city, the egotistical monadization that Engels calls the "fundamental principle" of capitalist society as a whole.[15] But Williams also brings out the possibilities of fulfillment offered by the city, and finally aspires not to a return to the rural but to a transcendence of the very opposition between city and country, in a post-capitalist society rooted in pre-capitalist values. Williams points out that this ideal was espoused by Marx and Engels, but was little developed in later revolutionary thought;[16] he also reminds us that Blake had already created a striking image of it, in his famous poem that advocates struggling to build "Jerusalem / In England's green and pleasant land."[17]

Towards socialism

In Williams's last period – the 1980s – while continuing to delve into authors he had earlier studied his work becomes increasingly aimed at defining the form that a "New Jerusalem" might take. At the same time, he reviews his past work in the sociology and history of culture, delineating the broad outlines of a general approach that he calls "cultural materialism."[18] We will mention here two works that illustrate on the one hand the further development of earlier subjects, and on the other the new orientation toward the future. The first is his monograph on William Cobbett,[19] an author he had already given attention to in both *Culture and Society* and *The Country and the City*, and who is, among the figures he had dealt with before, one of the closest to his own sensibility and point of view.

Although Williams acknowledges certain weaknesses in Cobbett, he considers them secondary. What makes Cobbett valuable as a writer, in Williams's eyes, is that he is profoundly democratic in temperament, that he has a highly concrete, spontaneously materialist mode of thought, and above all that he fully identifies himself with the workers in their suffering and revolt at the coming of capitalism. This is precisely the form of romanticism of which Williams is a proponent. In his study of the author of *Rural Rides*, [20] Williams emphasizes Cobbett's subversiveness against those who would see him as a simple reactionary:

> If he is the voice of anything that can be called, for rhetorical purposes, Old England, he is also in that act the voice of protest against finance capital, imperialism and the aristocratic State, and the voice of encouragement of working-class organisation, democratic protest and popular defiance.[21]

Although Williams considers Cobbett's thought to be in many ways surprisingly topical – notably in his early expression of ecological concern – in one respect he feels the early nineteenth-century polemicist is unable to be of help as the twenty-first century approaches. For his conception of production is overly narrow.[22] In Williams's view this notion must be rethought, and two years after his *Cobbett* he began to lay the groundwork in an article entitled "Towards Many Socialisms" (1985),[23] advocating a break with the categories of bourgeois economics. For bourgeois economics, only the "product" is important; the deleterious effects of production on humans and nature are only an unessential "by-product." For Williams, on the contrary, what is "produced" is in fact the overall society; product and by-product alike are part of production. What capitalism "produces" therefore includes the resulting human misery and environmental damage. Socialism must aim instead at the "production" of the most widespread possible human and natural well-being.[24]

While this concept is in some ways a new one, it is in a real sense also a return to a mode of thought preceding capitalism, a system of production that separates the economy – as the production of market goods – from the social totality, and elevates it to the rank of supreme value. And in his propositions for a renewal of the socialist project, Williams wants to revive the past as much as to innovate. He is convinced, it is true, that we will never return to the small, undifferentiated community, and he knows that a future socialist society must be complex, pluralistic, and technologically sophisticated. But for Williams what all of these appropriations of modernity should make possible is a greater fulfillment of the dreams and ideals of the past.[25]

In "Towards Many Socialisms," as in other essays of the 1980s, Williams attempts to bring together a socialism based on class analysis and the so-called "new social movements." We might see this as part of his larger project – formulated long ago – to reconcile Marxism and romanticism. Williams himself seems to recognize the relationship when in "Towards Many Socialisms" he points to the affinity between movements like ecology, feminism, and pacifism, and the romantic tradition that he has so finely interpreted.

The romantic tradition and nature

We will now examine Williams's intellectual trajectory more specifically from the ecological point of view – his attempts to deal with cultural attitudes towards nature on the one hand and threats to the environment on the other, first in his cultural studies and then in his directly political essays.

The issue of how the English romantic tradition relates to nature is not much developed by Williams in *Culture and Society*, but in *The Country and the City* we find some interesting comments. The most significant is a remark concerning the "baffling poem" *The Deserted Village* (1769) by Oliver Goldsmith, which leads to a general assessment of the question:

> Here, with unusual precision, what we can later call a Romantic structure of feeling – the assertion of nature against industry and of poetry against trade ... is projected. We can catch its echoes, exactly, in Blake, in Wordsworth, and in Shelley.[26]

Unfortunately, for some unexplained reason Williams does not deal, in this seminal book of 1973, with the confrontation of nature with industry in the works of the three poets mentioned. But the issue is discussed in relation to several other British romantic authors of the nineteenth and twentieth centuries.

The first one analyzed, in chronological order, is the poet John Clare – mentioned in our introduction – whose early poem, *Helpstone* (1809), is according to Williams "a familiar rural elegy and retrospect." It represents a typically romantic structure of feeling – what capitalist wealth "is most visibly destroying is 'Nature': that complex of land as it was, in the past and in childhood." Interestingly, Williams introduces here a very personal reflection concerning his own childhood and his own feeling of "loss of 'Nature'": "Over a century and a half I can recognise what Clare is describing: particular trees, and a particular brook, by which I played as a child, have gone in just this way ..." While distancing himself from Clare's contemplative and resigned attitude, Williams is attracted to his "green language" and his "natural images of the Eden of childhood;" "its strength is the connecting feeling of human warmth and community, in a time of real dispossession, eviction and social division."[27] As we will see later, childhood memories and Welsh community life are key components of Williams's own "structure of feeling."

Another poet and novelist mentioned in this context is George Meredith (1828–1909). His pastoral poetry shows a keen interest in nature as a place of physical and spiritual regeneration as well as of the "seasonal rhythm of the fundamental life process." However, Williams adds:

> neither of these feelings was new in itself. What was new was their fusion into a structure of feeling in which the earth and its creatures – animals and peasants almost alike – were an affirmation of vitality and of the possibility of rest in conscious contrast with the mechanical order, the artificial routines, of the cities.[28]

We are not convinced that this romantic structure of feeling was a "new" cultural form created by Meredith in the late nineteenth century. It has much earlier manifestations, particularly in England. But Williams's comment documents his interest in this "ecological" aspect of the romantic tradition.

Moving into the twentieth century, Williams observes that criticism of the (capitalist) industrial system as "a wasteland of ugliness and emptiness" can be made from a sort of "primitivist" perspective, as in D.H. Lawrence, who admired the Indians of New Mexico as well as all forms of "direct living in contact with natural processes – animals and birds and flowers and trees but also the human body ..."[29] A similarly romantic state of mind can also be found, according to Williams, in George Orwell's novel *Coming up for Air* (1939), which returns to "a vision of the country, the old unspoiled country, as a place of human retreat and rest, an innocence which the new civilisation, capitalist or socialist, was aggressively destroying."[30]

Curiously, although there are many pages on Charles Dickens in *The Country and the City*, his criticism of the industrial destruction of nature is hardly mentioned. The same may be said of several other authors belonging in one way or another to the romantic tradition, who are analyzed in the book. There is, on the other hand, a discussion of the attitude towards nature of William Morris, the thinker who plays a decisive role – as we saw above – in Williams's narrative of the transition from this tradition to socialism. Williams had already quoted, in *Culture and Society*, a striking passage from "Art and Socialism" (1884), in which Morris denounces the destructive work of (capitalist) commerce:

> less lucky than King Midas, our green fields and clear waters, nay the very air we breathe, are turned not to gold … but to dirt; and to speak plainly we know full well that under the present gospel of Capital not only there is no hope of bettering it, but that things grow worse year by year, day by day.[31]

However, at that stage of his intellectual evolution Williams did not really grasp the significance of Morris's powerful "ecological" admonitions, and his comment on them curiously misses the point: "For indeed," Morris argues, "the commercial habits of the middle class can destroy even those things which many individual members of the middle-class value."[32]

In the same 1958 publication, Williams is very critical of Morris's "regressive element," or his "medievalism." One may or may not agree with this critique, but it leads Williams to a serious mistake: leaving out of the list of Morris's significant writings his masterwork, *News from Nowhere* (1890), which Williams describes, along with Morris's "romantic socialist songs," as an example of his "disabling weaknesses." Somewhat in contradiction with this, Williams makes an insightful comment on the function of the "regressive elements" in Morris's utopia: "although their reference is to the past, their concern is with the present and the future."[33]

Williams briefly returns to Morris's *News from Nowhere* in *The Country and the City*, taking a more open, though still critical approach: "This is a combination of what is essentially restoration, turning back history and drawing on medieval and rural patterns … and a projected new London, in the contemporary sense of the garden city." Williams doesn't explain what he means by "garden city," but his general assessment of the utopian novel is ultimately a favorable one. Morris was able to tap into the energies of angry rejection of the modern capitalist order, and in his book "the negative energy has found a positive cause."[34]

As we saw earlier, Williams strongly rejects what he considers to be reactionary pastorals, hostile to democracy and to the labor movement. But this does not mean that he is hostile to the romantic structure of feeling concerning nature, quite the contrary:

> The song of the land, the song of rural labour, the song of delight in the many forms of life with which we all share our physical world, is too important and too moving to be tamely given up, in an embittered betrayal, to the confident enemies of all significant and actual independence and renewal.[35]

This "song of the land" was perhaps the inspiration behind the conclusion of *The Country and the City* – several pages that constitute a powerful anti-capitalist manifesto, in which the issue of damage to the environment is for the first time strongly emphasized.

Williams begins his argument by paying a direct homage to "some rural writers" (obviously belonging to the romantic tradition):

> One of the real merits of some rural writers, often not seen because other elements are present, is an insistence on the complexity of the living natural environment. Now that the dangers to the environment have come more clearly into view, our ideas, once again, have to shift.[36]

What does this shift mean? To move beyond, or better, to reject the "insane over-confidence" in metropolitan industrialism that has created the present situation, where "the risk to human survival is becoming evident;" in other words, to understand "the clear impossibility of continuing as we are." Moreover this holds not only for the city, and manufacturing industry, but for the country as well, considering "how much damage to the environment was and is being done by the capitalist mode of progressive agriculture."[37]

From this somber diagnostic Williams draws a sharp political conclusion, which is a striking appeal for (revolutionary) anti-capitalist action:

> We begin to see, in fact, that the active powers of minority capital, in all its possible forms, are our most active enemies, and that they will have to be not just persuaded but defeated and superseded … I am then of course convinced that resistance to capitalism is the decisive form of the necessary human defence.[38]

This impressive political statement, at the end of what is still a scholarly work, already announces the topics that will be dealt with in Williams's political and ecological essays of the 1980s.

Against the "endless drive for expansion"

Williams's interest in the romantic cultural tradition is probably one of the reasons why he became the first British socialist to engage, from a political perspective, with ecological issues. In *Towards 2000* (1983), in a chapter curiously entitled "East-West, North-South," he puts forward one of his key arguments. Capitalism follows the imperatives of the market, i.e. "the endless drive for expansion and accumulation," and the economies which claim to have superseded market imperatives – those of "actually existing socialism" – do the same, faithfully reproducing "the drives to indefinitely expanded production." The problem is that an infinity of expansion is impossible in a physically finite world.

Williams calls this "the ecological argument," emphasizing that it should not be reduced to a collection of specific issues: "the dangerous rising scale of industrial and chemical pollution, the destruction of some natural habitats and species." It is rather the present process of production itself that has to be challenged, and the worldview to which it corresponds:

> What is really at issue is a version of the earth and its life forms as extractable and consumable wealth. What is seen is not the sources and resources of many forms of life but everything, including people, as available raw material, to be appropriated and transformed.[39]

Williams is persuaded that one cannot continue to treat "the complex physical world and its intricate interacting biological processes" in this way without provoking "grave and unforeseen kinds of damage." One must take seriously, even if one doesn't necessarily accept all particular estimates, "the warnings of certain absolute kinds of damage," among which he mentions harm to "the atmosphere, to the climatic system," extreme shortages of necessary resources, "absolutely unacceptable levels of pollution and environmental damage," and widespread famine. The ambition of unlimited growth, of "indefinitely expanded production," is a dead end. Looking towards the year 2000, Williams insists it is impossible that the vast human population of the twenty-first century can be supported "at the currently wasteful levels of food and energy use, of the proliferation of short-duration and throwaway commodities, and of high artificial-input food production, now registered and even planned to rise in the most aggressive market economies."[40]

Williams does not develop the argument further, but his insightful remarks already raise most of the burning issues of the twenty-first-century ecological debate, including climate change. His conclusion is not particularly optimistic, but is rather a sober reminder of the difficulties on the road to the needed social-ecological, radical transformation, and the dangers involved in the pursuit of the present course of development:

> The changes involved are so substantial and resistance from existing interests so certain and powerful, that nobody can suppose that this will be anything but a very long and complex struggle. The issues have then to be restated in their hardest terms. It is already a distinct possibility that the most aggressive marketing and militarist societies will destroy the world simply by continuing their own current policies.[41]

Thirty-four years later, we are still in the middle of this "very long and complex struggle," and the risk of a world catastrophe is still with us. Raymond Williams was one of the few, in England and elsewhere, who warned of these developments in the 1980s.

In *Towards 2000* Williams gives only a few first suggestions in the search for an alternative, a different way of thinking that would be able to break radically with the "basic orientation to the world as available raw material," which is shared by both the capitalist economy "in its relentless drives for profit and accumulation" and the so-called "communist" or "actually existing socialist" economies.[42] Williams asserts the need to combine, or better, to unite, the ecological with the political and economic arguments developed by the movements for democracy and socialism. To do so is not, he recognizes, an easy task. It requires "a new kind of politics" and a deep change of mind, not only rational but also "emotional." Only such a change could generate "the energies and the practical means of an alternative social order." The central element of such an order, Williams believes, would be "the shift from 'production' to 'livelihood,' from an alienated generality to direct and practical ways of life."[43] He developed a more specific and concrete version of this new politics in his last great book, *Resources of Hope* (1989), to which we now turn.

For a socialist ecology

In 1973 a small leftist ecological group, the Socialist Environment and Resources Association (SERA), was founded in England. An independent environmental association, it was however affiliated with the UK Labour Party. When SERA invited Williams to become their vice-president he gladly accepted. In the autumn of 1981 he gave a talk for them on "Ecology and Socialism," which was soon published as a pamphlet and later (posthumously) reprinted in *Resources of Hope.* [44] This pioneering article was among the first in Europe to associate these two terms, from a (heterodox) Marxist perspective.

Williams begins by saying that "in recent years some of us have been talking about ecological socialism" – a statement that probably refers to his friends from SERA. After paying due homage to the German biologist Haeckel, who invented the concept of ecology in the 1860s, he discusses the first reactions to the dire consequences of the Industrial Revolution for the environment. The first example he mentions is a striking statement, given its source, by James Nasmyth, the inventor of the steam hammer, around 1830: "The grass has been parched and killed by the vapours of sulphureous acid thrown by the chimneys, and every herbaceous object was of a ghastly grey – the emblem of vegetable death in its saddest aspect. Vulcan had driven out Ceres."[45]

The German socialist author Hans-Magnus Enzensberger had written, in an article published by *New Left Review* in 1974, that pessimistic views on the environmental harm caused by industrialization appeared only when the effects reached districts where the bourgeoisie lived. Williams sharply disagreed:

> Now this is simply untrue. From Blake and Southey and Cobbett, in the early decades of industrialization, through to Carlyle, Ruskin, Dickens and William Morris, there were constant observations and arguments just of this kind. I analyzed many of them in *Culture and Society*. It remains a curiosity that this whole body of social observation and argument, which arose very early in Britain for the obvious reason that the most spectacular industrialization was taking place here, is often not known at all by educated continental socialists, who can then build a wholly misleading history of ideas.[46]

As we have seen above, environmental issues are not very developed in *Culture and Society*. It would have been more accurate to mention *The*

Country and the City. But the main point is that looking backwards, from his present socialist-ecological viewpoint, Williams clearly perceives "this whole body of social observation and argument" – the British romantic critics of capitalist civilization, from Blake to William Morris – as the forerunner of the new perspective. A whole section of the lecture is devoted to Morris, whom he praises as the first to begin to "unite these diverse traditions, in British social thought" – the romantic and the socialist. The great contribution of Morris is his criticism of both productivism and consumerism:

> His critique of the abstract idea of production was one of the most decisive interventions in the socialist argument. Instead of the simple capitalist quantum of production, he began asking questions about what kinds of production. In this, in fact, he was following Ruskin, who argued much the same case – who insisted that human production, unless governed by general human standards, rather than mere profit or convenience, could lead to "illth" as readily as to "wealth" ... [Morris] sees human need as something more than consumption, that incredibly popular idea of our time, which from the dominance of capitalist marketing and advertising tries to reduce all human need and desire to consumption.[47]

Williams then opposes Morris's sharp insights to what he calls "weak socialism": a form of socialist thinking in which, as much as in "ordinary capitalist thinking," "production becomes insensibly an end in itself." This criticism of productivist socialism – probably meaning the dominant current in the Labour Party – is an important aspect of Williams's argument. He explains it by "capitalist influence":

> Under capitalist and imperialist influence, and especially since 1945 under North American influence, the majority position among socialists has been that the answer to poverty, the sufficient and only answer, is to increase production. This in spite of the fact that a century and a half of dramatically increased production, though it has transformed and in general improved our conditions, has not abolished poverty, and has even created new kinds of poverty, just as certain kinds of development create underdevelopment in other societies.[48]

Williams acknowledges however that this form of thinking existed well before the "North American influence," and has its roots "as far back as Bacon's *New Atlantis*," in which the aim was the "conquest of nature." Unfortunately, this idea of "conquest" or "mastery" of nature appears "not only in the dominant bourgeois thought but also all through socialist and Marxist writing in the second half of the nineteenth century." Among the exceptions is Friedrich Engels, who, in a passage from *Dialectics of Nature*, insists that "we are ourselves part of nature" and therefore "can't just arrive and depart as a foreign conqueror." And today we still read such triumphant statements about conquering nature and expanding production, in the name of socialism.[49]

This is a powerful argument, but it is somewhat limited by the emphasis on "forms of thinking," with little discussion of their practical consequences in concrete economic life. That limitation leads to another surprising silence with regard to the triumphant and reckless productivist strategy, in the name of "communism," of the Soviet Union. Williams was an outspoken critic of Stalinism, and, as we saw above in *Towards 2000*, of the Soviet relation to nature as "raw material." But nowhere in the 1981 lecture does Williams deal with the consequences for the environment of "really existing socialism."

Williams's criticism is not only directed against "weak socialism" but also "weak ecology," or what he calls "non-political ecology." It is not clear if he is referring here to non-political ecological movements in general, or more specifically to the British Ecological Party (founded in 1975). In any case, he criticizes their rejection of the Right-Left divide, and above all their illusions about "leaders of the world." He has no objection to the concrete proposals of the ecologists – to reduce energy consumption, to outlaw certain harmful manufacturing processes, etc. The problem is that the non-political ecological approach calls on "world leaders" to implement these measures. In other words, it is "calling upon them to go against the precise interests, the precise social relationships, which have produced their leadership." This political position, Williams thinks, "can be worse than merely mistaken, because it creates and supports the notion that the leaders can solve these problems." Nothing could be further from the truth: "Increasingly … the effective ruling class dismisses the whole case [of ecology] as sentimental nonsense, which simply limits or delays production and national power."[50]

Beyond "weak socialism" and "non-political ecology," Williams calls for an "ecologically conscious socialism" able to understand the limits of the planet and the impossibility of unlimited productive growth. This, as he sees it, is not a matter of choice. We cannot go on with the "existing patterns and conditions of production, with all their actual looting of the resources of the earth and all their damage to life and health."[51]

"Walking Backwards into the Future"

We should address, finally, another critical commentary made in Williams's lecture of 1981. It concerns what he considers to be a weakness of the romantic tradition – including the work of William Morris – as well as of the present ecological movement: "an inbuilt tendency to contrast the damaging industrial order with the undamaging, natural, pre-industrial order." Arguing against this perspective, Williams asserts that disasters such as the destruction of forests are occurrences that existed well before industrialism, and that such devastation "probably goes back to Neolithic times."[52] This affirmation, even in 1981 when awareness of the extent of the damage caused by modern industrial capitalism was less acute than it is today, seems strange indeed. How can one compare Neolithic times with contemporary capitalist modernity in terms of consequences for the environment? It should be added that, in contrast to this perspective, in the first volume of Williams's novel *People of the Black Mountains* (posthumously published in 1989), several chapters describe the way of life of this Welsh region in the Neolithic era. Far from being "destructive," these communities of hunters and shepherds are seen as peacefully living in harmony with their natural environment.

Later in the lecture, and in the same perspective, Williams aims a sharp criticism at William Morris, claiming that he was a victim of the delusion that:

> before factory production, before industrial and mechanical production, there had been a natural, clean, simple order. For Morris this ideal order was located, as for so many nineteenth-century radicals and socialists, in the Middle Ages. Thus a notion that the future, the socialist future, would be some kind of reconstitution of the medieval world, established itself deeply in his thought, although it always worried him.[53]

This, for Williams, is an illusion that survived long after Morris: "I don't have to tell you how that kind of thinking still is within the ecological movement."[54] But to reduce Morris's utopian image of the socialist future to a "reconstitution of the medieval world" is to paint a rather inaccurate picture of the subtle dialectics between past and future in *News from Nowhere*. A return to medieval times would mean the restoration of the absolute power of King and Church, as well as of feudal social relations, peasant serfdom, and religious intolerance – all of which are of course entirely absent from Morris's anarcho-communist utopia. As in *Culture and Society*, in spite of his admiration for Morris, Williams seems here unable to grasp the greatness of his most important piece of writing.

This sort of argument might give the impression that Williams was moving away from the romantic tradition in the early 1980s when he gave his lecture. But elsewhere he reaffirms it, in a more personal vein, as we mentioned above. In an article entitled "Between Country and City" (1984), he enumerates, with obvious sympathy, "the occupations of my neighbours within five miles of my house in the Black Mountains": "weavers and knitters; potters; cabinet-maker; pine furniture maker; booksellers; book-illustrator; antique-clock restorer; antique dealers; painter and gallery owner; writers; sculptor; restaurateurs; glass engraver; stained glass maker." An interesting list, which has a striking similarity to the crafts cherished by William Morris, and mentioned in *News from Nowhere*. Williams then adds a significant comment: "The majority of all these are comparatively recent immigrants, but consider how much they are, taken as a whole, restoring a genuine fabric of rural society." The key word here is *restoring* – a typical romantic concept.[55]

A few lines later Williams celebrates the way people living now in the Black Mountains practice the "recovery, exploration and propagation of kinds of natural knowledge ... which were in part drained out of the surviving rural population in the period of urban and industrial domination." They are "interacting with the best of the surviving rural economy: in the wholefoods, the honeys, the culinary herbs, the fuels, the jams and cordials." All this can be described as "a healthy practical recovery of the skills and resources of the land." In this passage, the key romantic term is *recovery.* [56] What is involved is not "regression" but restoring and recovering past traditions of arts and crafts, and of the relation to nature. Thus, when writing about the Black Mountains Williams is at his nearest to Morris. The issue is again raised in an essay from 1985, with the suggestive title "Walking Backwards into the Future." This romantic proposition is not really discussed in the article, but in the first paragraph we find a telling statement:

> Most people want to change our present social and economic conditions, but it is noticeable how many of the words we use to define our intentions have a reference to the past: recovery, rehabilitation, rebuilding. In fact ideas of a better past and of a better future have chased each other through all modern political thought. Many early radicals believed that there had been a better and happier time, just before recent disastrous changes, and that was what they had to recover or restore.[57]

In this case Williams does not reject that way of thinking and, as we saw, he himself uses these same words when speaking of the Black Mountains.

What then is for Williams the alternative solution? To stop considering "the earth and its life forms as raw material." Against capitalism, industrial and rural, and unlike productivist socialism, "the most hopeful social and political movement of our time" is the "now emergent green socialism," able to unite ecology and economics in a "new politics of equitable livelihood."[58] While new and modern, this *green socialist* tree clearly also has deep roots in the romantic critique of capitalist/industrial civilization.

Notes

1 May Day Manifesto, 45.

2 For example, in a review of Michael Löwy's *Georg Lukács: From Romanticism to Bolshevism*, Williams refuses to see Bolshevism as a simple "progress" in relation to the diffuse anti-capitalist romanticism of the young Lukács and others. He points out that since the romantics denounced state bureaucracy, the quantification of thought under industrialism, and the lack of community in modern society, "we can hardly, from the end of the seventies, suppose that they were wasting their time or missing some simple central truth" (*New Society*, 24 January 1980, 189).

3 Williams's novels, which are set in the region of Wales where he was raised, also exhibit romantic traits. Notably, his last, posthumously-published novel – *People of the Black Mountains* – returns to the prehistory of the region, portraying early

hunting communities that are unified and close to nature. Their mode of life is shown as limited, but rich in terms of the quality of their human and natural bonds.

4 For monographs on the whole of Williams's work that stress its continuities, see Gorak, *The Alien Mind of Raymond Williams*, and Inglis, *Raymond Williams*.
5 Williams, *Politics and Letters: Interviews with New Left Review*, 97–98.
6 Williams, *Culture and Society, 1780–1950*, 161.
7 Williams, *Culture and Society*, 280.
8 Williams, *Culture and Society*, 328.
9 Williams, *Politics and Letters*, 106.
10 Williams, *Politics and Letters*, 115.
11 Williams, *Politics and Letters*, 314.
12 Williams, *Politics and Letters*, 304, 314.
13 On the general phenomenon, see Williams, *The Country and the City*, Chap. 2.
14 On the notion of "knowable communities," see Williams, *The Country and the City*, Chap. 16.
15 See Williams, *The Country and the City*, 150, 215–216.
16 Williams, *The Country and the City*, 304.
17 Williams, *The Country and the City*, 149.
18 See in particular Williams, "The Uses of Cultural Theory."
19 William Cobbett (1763–1835) was famous in his time as a journalist, essayist, and polemicist. After early phases of republicanism followed by conservatism, he became a "radical" after 1804.
20 This best-known work by Cobbett has achieved the status of a minor classic: see Cobbett, *Rural Rides*.
21 Williams, *Cobbett*.
22 Williams, *Cobbett*, 79.
23 The essay has been republished in the posthumous anthology: Williams, *Resources of Hope: Culture, Democracy, Socialism*. In this anthology are to be found most of Williams's articles and papers of the 1980s. He develops his ideas for a socialist future at greater length in *Towards 2000*.
24 Williams, *Resources of Hope*, 308–309.
25 See, for example, Williams, *Resources of Hope*, 306.
26 Williams, *The Country and the City*, 79.
27 Williams, *The Country and the City*, 138–140.
28 Williams, *The Country and the City*, 252.
29 Williams, *The Country and the City*, 266.
30 Williams, *The Country and the City*, 275.
31 Williams, *Culture and Society*, 156.
32 Williams, *Culture and Society*, 156.
33 Williams, *Culture and Society*, 159.
34 Williams, *The Country and the City*, 273.
35 Williams, *The Country and the City*, 271.
36 Williams, *The Country and the City*, 301.
37 Williams, *The Country and the City*, 301.
38 Williams, *The Country and the City*, 301–302.
39 Williams, *Towards 2000*, 215.
40 Williams, *Towards 2000*, 215–216.
41 Williams, *Towards 2000*, 217.
42 Williams, *Towards 2000*, 261. Williams argues that Marx himself "shared with his capitalist enemies an open triumphalism in the transformation of nature, from the basic orientation to it as raw material" (264). This is a much too one-sided assessment, one which does not take into account the complexity of Marx's

relation to nature. For the opposite emphasis, see Foster, *Marx's Ecology: Materialism and Nature*.
43 Williams, *Towards 2000*, 263, 266, 267.
44 Inglis, *Raymond Williams*, 273.
45 Williams, *Resources of Hope*, 211.
46 Williams, *Resources of Hope*, 212. A few years later, in the article "Towards Many Socialisms" (1985), Williams again emphasizes the link between what he calls "anticapitalism" (no authors mentioned) – a critique of the bourgeois industrial social order rejected by "unilinear socialism" as "an immature or romantic stage" – and the contemporary ecological movement: Williams, *Resources of Hope*, 306–307.
47 Williams, *Resources of Hope*, 216.
48 Williams, *Resources of Hope*, 215.
49 Williams, *Resources of Hope*, 214.
50 Williams, *Resources of Hope*, 219.
51 Williams, *Resources of Hope*, 221, 225.
52 Williams, *Resources of Hope*, 212.
53 Williams, *Resources of Hope*, 216–217.
54 Williams, *Resources of Hope*, 216–217.
55 Williams, *Resources of Hope*, 214.
56 Williams, *Resources of Hope*, 234–235.
57 Williams, *Resources of Hope*, 281.
58 Williams, *Resources of Hope*, 237.

Works Cited

Cobbett, William. *Rural Rides*. Harmondsworth: Penguin, 1985.

Foster, John Bellamy. *Marx's Ecology: Materialism and Nature*. New York: Monthly Review Press, 2001.

Gorak, Jan. *The Alien Mind of Raymond Williams*. Columbia, MO: University of Missouri Press, 1988.

Inglis, Fred. *Raymond Williams*. London: Routledge, 1995.

Löwy, Michael. *Georg Lukács: From Romanticism to Bolshevism*. London: New Left Books, 1979.

May Day Manifesto. Edited by RaymondWilliams. Harmondsworth: Penguin, 1968.

Williams, Raymond. *Cobbett*. Oxford: Oxford University Press, 1983.

Williams, Raymond. *The Country and the City*. New York: Oxford University Press, 1973.

Williams, Raymond. *Culture and Society, 1780–1950*. New York: Harper & Row, 1958.

Williams, Raymond. *People of the Black Mountains*. London: Chatto and Windus, 1989.

Williams, Raymond. *Politics and Letters: Interviews with New Left Review*. London: New Left Books/Verso, 1979.

Williams, Raymond. *Resources of Hope: Culture, Democracy, Socialism*. Edited by Robin Gable. London: Verso, 1989.

Williams, Raymond. *Towards 2000*. London: Chatto and Windus, 1983.

Williams, Raymond. "The Uses of Cultural Theory." *New Left Review* 1, no. 158 (1986): 19–27.

6 Naomi Klein, twenty-first century climate warrior

The first chapter of this book dealt with an exemplary figure of the eighteenth century, William Bartram, whose romanticism expressed an ecological awareness that was quite unusual for that early period in the development of capitalist modernity, and that defined him as a prescient forerunner. In this last chapter, we take up a writer and activist who is still practicing today, and who forcefully demonstrates the continuing presence – and power – of the conjunction of romanticism and ecology in our contemporary world. Naomi Klein differs in several significant ways from the subjects of our previous chapters. She is of course first of all a woman, whereas our earlier figures are male, a fact that reflects the sharply growing trend since the late twentieth century for women to enter and excel in areas previously dominated by men. But also, it should be noticed, as a journalist Naomi Klein comes to romanticism following a somewhat different path from that of the others, one that is less literary, less directly linked to romanticism as a cultural tradition. In spite of this latter difference, however, she is very much a part of the same strand that we have been exploring, which joins a romantic anti-capitalist sensibility with a primordial concern for the fate of the natural world.

Klein's journalistic career began in the early 1990s, but her major contribution has been the publication of a series of important books that started with *No Logo*, first published in 1999. *No Logo* analyzed the phenomenon of globalized corporate capitalism and the movements of resistance to it, and was followed several years later by a collection of her newspaper articles and talks dealing with related themes: *Fences and Windows* (2002). But her next major work, which introduced an entirely new analytical framework for the understanding of contemporary capitalism, was *The Shock Doctrine* (2007). After a considerable hiatus of time *This Changes Everything: Capitalism vs the Climate* (2014) appeared, again constituting a new departure, for the first time focusing on the environmental dimension of the contemporary "globalized" world. Most recently, Klein has published *No is Not Enough: Defeating the New Shock Politics* (2017). Written as an urgent response to the victory of Donald Trump and related international developments, it integrates all of the analytic concepts and concerns of the earlier books.

While the writings of Klein, from *No Logo* to *No is Not Enough*, show a strong continuity in overall vision and sensibility, they also undergo a clear evolution in several respects. First, Klein's socio-political position becomes markedly more radical in her later books, though not in any sectarian fashion. From the beginning Klein developed a through-going and severe critique of the most recent forms taken by the capitalist system, explicitly identified as such. But in both *No Logo* and *The Shock Doctrine* she stops short of calling for far-reaching, or revolutionary social transformation, and in the introduction to the latter seems to identify herself with the Keynesian perspective.[1] In the last two books, by contrast, Klein aspires to more radical change. The main thrust of the argument of *No is Not Enough* is, as the title suggests, that opposition to things as they are will not suffice and that it is necessary to articulate a vision of an alternative society; to do so, Klein claims, we must engage in utopian thinking, dare to dream and to take a "leap."[2]

The other important development in Klein's work is a shift in focus towards the ecological crisis, and in conjunction with it, towards concern with indigenous peoples. The point at which she turned her attention primarily to ecology, and more specifically climate change, can be dated to the period around 2009, when she joined the governing board of the anti-climate change group 350.org, and the refocus begun then culminated in *This Changes Everything*. We will argue that it is through this shift towards ecological concern, and at the same time concern for the fate of indigenous peoples, that Klein's vision fully coalesces into a form of romantic anti-capitalism, and we will discuss *This Changes Everything* in this light. But there are significant elements of the romantic vision already present in the earlier writings, and before turning to *This Changes Everything* we will look briefly at them.

No Logo and *The Shock Doctrine*: "branding" the present and "blanking" the past

In *No Logo* Klein discusses with a wealth of detail the set of developments that takes place starting in the mid-1980s in advanced capitalist countries, particularly the United States, and that amounts to a paradigm change. The principal shift is from a regime in which companies still directly make products to one in which in many cases actual production is outsourced, and the company's major activity comes to be "branding." In this situation advertising passes to a new level, becoming the main activity, and source of profitability, of the company. In her analysis Klein shows how branding, like capitalism more generally, has an infinitely expansive dynamic, and how its specific characteristic is to generate a spurious association of the brand with certain non-quantitative "values," fabricating illusory identifications of the marketed product with ethics, spirituality, magic, and the like. In one striking passage, she calls brands "transcendent meaning machines."[3] Although Klein does not explicitly thematize the connection, these kinds of non-quantitative values are ones that played important roles in pre-capitalist societies and were largely replaced by the dominance of quantitative value under capitalism. Consequently, the brand can be seen as a wholly inauthentic, ersatz answer, from within a thoroughly commercial framework, to the yearning for values incarnated in some form in past societies – a yearning at the heart of romanticism.[4]

Klein's second major book, *The Shock Doctrine*, conceptualizes and explores in depth another crucial aspect of contemporary society: the imposition, often by force, of that extreme form of deregulated, fully privatized, "fundamentalist" capitalism that is usually referred to as neoliberalism. As propounded by the "Chicago School" theorist Milton Friedman, this doctrine is difficult to implement without the help of some kind of shock – natural or manmade – that disorients and weakens the social forces that might otherwise oppose it. Klein documents how this purist version of unbridled capitalism, which eliminates all other considerations than the free play of market forces, has in the recent past taken advantage of moments of collective trauma such as a political coup, a war, a tsunami, or a stock market crash, to impose its agenda. In this process, as she describes it, there is a concerted, aggressive effort to entirely wipe clean the slate, to erase or "blank" all that came before – that is, the past.[5] In this way all traditional cultures, practices, and institutions are negated and swept away, and new, pure market structures are put in their place. In her analysis of the "shock doctrine," then, Klein highlights the radical break with, and opposition to, the past that capitalism effects, although in these recent developments what is involved, of course, is usually only traces, or pockets, of pre-modern ways of living.

In the concluding sections of both *No Logo* and *The Shock Doctrine*, and throughout the journalistic texts anthologized in *Fences and Windows*, Klein reports and reflects on the movements of resistance of various kinds to contemporary capitalism that have grown up, expanded and diversified since the late 1990s. In her remarks on them we find elements already present of the romantic ecological perspective that will come together in a more coherent and fully developed way after her shift in orientation around 2009, articulated in *This Changes Everything*. Klein pinpoints as a key aspect of the protests against the big brand companies the effort to "de-fetishize commodities," and calls for the movement to go even further in "recogniz[ing] the need for non-commodified experiences and [reawakening] our desire for truly public spaces."[6] In discussing a number of different actions she uses the term "commons" to describe the latter, and although she does not explicitly refer to the historical enclosure of commons that was integral to the rise of capitalism, that often seems clearly implicit. Her discussion of the "Reclaim the Streets" movement, for instance, emphasizes that it aims to "take back" something – a public space – that had previously been collectively held and enjoyed. Though "Reclaim the Streets" is of course an urban movement, Klein relates it to ecological movements that have applied the same notion of commons to wilderness spaces.[7]

In the writings that precede *This Changes Everything* the issue of damage to the natural environment is not particularly foregrounded, and remains just one among many other negative consequences of globalized capitalist development. But occasionally Klein's commentaries show that she was already beginning to be aware of the importance of the question. In an article on NAFTA in 2001, for instance, Klein questions the prevailing doxa of growth at all cost, asking "Why can't we measure ecological deficits as well as economic growth?"[8] She also in a few cases mentions indigenous groups in relation to the problem of ecological destruction carried out by companies exploiting natural resources. In the final section of *No Logo*, dealing with resistance initiatives, she describes the campaign carried out by the Lubicon Cree Indians of Northern Alberta and a support group, against logging and mining operations that had "caused massive damage to the ecosystem and to the Lubicon way of life."[9]

One of the epigraphs for the concluding chapter of *The Shock Doctrine* is a quotation from Evo Morales's inaugural presidential address to his "Indian brothers," showing that Klein was also interested early on in wider pan-American indigenous struggles.[10] Even more indicative of an interest in indigenous peoples, and one that has unmistakable romantic overtones, is an article on the Chiapas and the Zapatista movement, first published in *The Guardian* in 2001. In it she speaks of how its leader, Subcomandante Marcos, is "consciously trying to appeal to something that exists outside the intellect, ... that he found in himself in the mountains of Chiapas: wonder, a suspension of disbelief, plus myth and magic." He, as an urban intellectual, had discovered those things in an "ancient culture" that was not in any way commercialized, but rather "genuine, anachronistic folklore."[11] In this passage, then, it is clear that already at the turn of the twenty-first century Klein was beginning to develop some of the interests and ideas that would only later come to take center stage. In *This Changes Everything*, to which we turn now, isolated, occasionally expressed fragments of a romantic ecological vision come together into a more focused, integrated whole.

This Changes Everything: Naomi Klein as climate warrior

In our view Klein's *This Changes Everything* is a superlative book, the importance of which far exceeds most writings on the issue of climate change. Rejecting the illusory politics of "polite piecemeal" improvements, it is a passionate plea to organize and fight for radical change – against the established order, for climate justice and at the same time for a new civilization. In this work, it would seem, the struggle to stop global warming has found an exemplary compass.

We might first address the question of how this remarkable work belongs to the cultural universe of romantic anti-capitalism. Unlike the other authors discussed in our study, Klein has little connection with the classic romantic tradition. For sure, there are a few references to the great romantic authors of the past. For example, she quotes William Blake's famous passage about the "dark Satanic Mills," adding that they "blackened England's skies" and thus were responsible for "the beginning of climate change."[12] She also refers to a wise comment by

Victor Hugo in 1840: "How sad to think that nature speaks and mankind doesn't listen" (29). There are as well a few mentions of American representatives of the romantic tradition, such as Henry David Thoreau and John Muir (one of the founders of the Sierra Club in 1892), who often drew heavily on Eastern beliefs and "Native American cosmologies that see all living creatures as our 'relations'." Klein points out that Muir went so far as to denounce "the devotees of raving commercialism" and their idolatry of the "Almighty Dollar" as a threat to nature (183–184). But these references are marginal in her text. And indeed, unlike so many nineteenth-century poets and artists, Klein does not see nature through the eyes of religious spirituality or an aesthetic sensibility. It might be asked whether romanticism in the twenty-first century does not have the same direct relation to the early romantics, as was still the case in the twentieth century. The question remains open, but this at least seems to be the case with Klein.

It should also be mentioned that, unlike the authors earlier discussed whose works date from the second half of the nineteenth century on – William Morris, Walter Benjamin and Raymond Williams – Klein's work has little connection to Marx and the Marxist tradition. There is no hostility to Marx in her book, and there is even a sympathetic quotation from a passage in *Capital* in which Marx refers to capitalism's "irreparable rift" with "the natural laws of life itself" (177). However, Klein's way of presenting the forces responsible for disastrous climate change is far from the Marxist approach to capitalism as a mode of production. She often seems to see the culprit as being mainly an ideology – market fundamentalism, free-trade orthodoxy, the extractivist mind-set (19, 25, 63, 86, 443, 459–460) – and the struggle for change as a "battle of world-views."

Of course ideology is important, and Klein's critique of the broad modern cultural context – defined as "Post-Enlightenment Western culture" – whose origins she situates in Francis Bacon's concept of "hounding" nature, is quite relevant. But putting the primary emphasis on ideas risks leading to a purely "idealist" view of the process, instead of a systemic "materialist" one. Fortunately, she does refer to the system, though again not quite in Marxist terms. For instance, she often speaks of the economic structure leading to environmental destruction as "un-regulated capitalism" (20, 24, 63). Would a "regulated capitalism" suffice to remedy the situation? She never says so, but some of her proposals – such as a "Marshall Plan for the climate" – could be so interpreted (458). In some passages Klein refers to the "rules of capitalism as they are currently constructed" (88), which again seems to leave open the possibility of a different set of capitalist rules, perhaps along neo-Keynesian lines.

However, there is no lack of forceful and explicit statements that seem to aim at capitalism as a whole in *This Changes Everything*. Notably, reflecting the subtitle of the book she defines the decisive conflict of our times as a "battle between capitalism and the planet" – one, she adds, that is already under way, though "right now capitalism is winning hands down" (22). In addition, there is the reiterated idea that the threat of irreversible environmental catastrophe presents a unique opportunity for breaking the status quo and allowing an opening for broad social transformation. Thus, although some ambiguity remains concerning the extensiveness of the change she foresees, and calls for, overall the book seems to represent a significantly more radical perspective than her previous ones.

Ancestral traditions and indigenous struggles

If romanticism is a cultural protest against modern capitalist civilization in the name of past values, the romantic dimension of Klein's book clearly has to do with her view of the indigenous struggles around the planet, in defense of life itself. She belongs, without necessarily being aware of it, to a long tradition of romantic/revolutionary – democratic, socialist, or anarchist – admirers of "primitive," "savage," or indigenous communities: Jean-Jacques Rousseau, William Bartram, Lewis Morgan, Friedrich Engels, Elisée Reclus, and Walter Benjamin (among others). However, unlike most of them her attitude is not just a nostalgic sympathy for past forms of life but an active commitment to presently struggling communities, similarly to other romantic radicals such as José Carlos Mariátegui, the Peruvian Marxist of the 1920s who celebrated the emancipatory struggle of the Andean indigenous groups, inheritors of "Inca communism;" or the Surrealists, who since the 1920s have shown not only a keen interest in non-Western cultures but also an active solidarity with the anti-colonial resistance of the "savages."

Klein understands, better than many traditional leftist movements blinded by a dogmatic "worker-ism," the decisive importance of indigenous struggles for the environment. They are, quite literally, in the front line of the war to save life on the planet. By protecting their beloved forests, mountains, rivers and coastlines they are, Klein asserts, "helping to save us" (352). One particularly important example is the struggle of the Amazonian indigenous groups. By holding back the oil interests determined to sacrifice new stretches of the forest, they are protecting the last great "carbon-capturing" sink of the planet (376). It is not surprising, therefore, that "many non-Native people are starting to realize that Indigenous rights … may now represent the most powerful barriers protecting all of us from a future of climate chaos" (380).

This sincere feeling of gratitude towards indigenous peoples is not in itself romantic. What gives a distinctive romantic "color" to Klein's commitment to these communities is her deep admiration for their ancient traditions, in relation to nature. For her, these pre-modern ways of life, whose origins lie in past centuries, are not only profoundly antagonistic to the modern destructive ethos of capitalism, but also illuminate the path towards a new future. In other words, her views clearly belong – in their own idiosyncratic way – to the revolutionary/utopian variant of romanticism.

Klein's emphasis is on the Americas, but she believes that Europe too had such a "naturalist" past, progressively destroyed by the ruling classes, particularly after the seventeenth century. In England, for instance, the British elites, following the views of Sir Francis Bacon, abandoned "once and for all, pagan notions of the earth as a life-giving mother figure to whom we owe respect and deference" (170). This older way of relating to the earth is the background for the rebellion, from the earliest days of industrialization, of English and Irish peasants against the enclosures of communal lands (177). And it inspires, in our time, the stubborn resistance of the villagers in Halkidiki, Greece, against the destructive project of the Canadian mining company Eldorado Gold to clear-cut large swaths of the old-growth Skouries forest and "reengineer" the local water system. One of the movement leaders, the young mother Melachrini Liakou, explained in an interview with Klein in 2013 the motives for this opposition, largely supported by Greek public opinion: "I am part of this land. I respect it. I love it and I don't treat it as a useless object ... In contrast, Eldorado, and any other mining company, they want to devour the land, to plunder it," leaving behind nothing but waste and "a huge chemical bomb for all mankind and nature" (296, 342).

In fact, Klein points out, this is a universal phenomenon that has a long history – centuries of indigenous resistance against colonizers around the world (177). And today "world views based on reciprocity and interconnection with the natural world" can be found in North America, Latin America, Nigeria, Australia and New Zealand, among others, and they play a key role in the present struggles against capitalist extractivism (182) – struggles which Klein designates by the striking term Blockadia. Even in China, under conditions of severe state repression, peasants who rely on traditional subsistence activities like small-scale agriculture and fishing "have a history of militant uprisings against industrial projects that cause displacement and disease, whether toxic factories, highways or mega-dams" (350–351).

Sometimes these movements are able to influence government policies. But such results are often short-lived. A striking example described by Klein is the Yasuni project, proposed by indigenous and ecological movements in Ecuador, to preserve the Yasuni Park rainforest – a unique haven of biodiversity – from the schemes of voracious oil companies to reap the millions of barrels of crude underneath the land. A call was to be issued for the international community to compensate the country for not extracting the oil, with the money raised – only half the value of the oil reserves – going to social programs and a clean energy transition. It would have created a precedent, showing that poor countries should be rewarded for leaving the oil in the soil – the best and surest way to reduce emissions of carbon dioxide. When elected in 2007, the Center-Left government of Rafael Correa adopted the Yasuni project, which had overwhelming popular support. A poll in 2011 found that 83% of Ecuadorians agreed to leave the oil in the ground. However, the contributions from developed countries did not arrive (or only in very small amounts), since the project did not fit into their cost-effective market calculations. Unfortunately, Rafael Correa decided in 2013 to allow drilling to begin. For Klein, this failure illustrates the contradictions of "progressive" governments, but above all the unwillingness of the "international community" – as shown also by the broader failure of the climate negotiations – to take the serious measures needed to reduce emissions. Indigenous movements in Ecuador, and various citizen groups, have protested and collected 750,000 signatures calling for the matter to be put to a national referendum, and the struggle continues (408–411).

Born and living in Canada, Klein has an especially keen interest in the Canadian First Nations, who have been the most decided and obstinate adversaries of the extraction – and pipeline transportation – of tar sand oil, the dirtiest of all fossil energy sources on the planet. These are communities whose customs have existed for "hundreds of generations;" their roots in what is now western Canada "go back some ten thousand years." These roots in the past are not only important for elders. The young people living in these communities know very well that "their health and identity [are] inextricably bound up in their ability to follow in the footsteps of their forebears" – i.e. to fish in the same waters, hunt in the same forests, and collect medicines in the same meadows. This is why the British Columbia indigenous peoples saw the building of the Northern Gateway Pipeline – a gigantic enterprise designed to transport huge quantities of toxic tar sand oil – as a brutal "wave of colonial violence" leading to the destruction of "the lands and waters that sustained our ancestors" (299, 309, 341–342).

A word that is often used by Klein to describe the ancestral practices of these communities is "stewardship." According to the Merriam Webster dictionary, this means "careful and responsible management of something entrusted to one's care." This definition perfectly fits the way First Nations relate to nature – one has only to add "entrusted to one's care by several generations of forebears." This stewardship has a sort of special affinity with the sun, as an alternative not only to fossil fuels but more generally to the modern Western artificial lifestyle. As Red Cloud, a Lakota Sioux leader engaged in renewable energy projects in the western United States, expresses it: "solar power was always part of Natives' way of life. Everything followed the *anpetuwi tawonawaka*, the life-giving force of the sun. It ties with our culture, our ceremony, our language, our songs" (393). Similar words are used by Vanessa Braided Hair, a young Cheyenne activist. Solar power embodies the worldview in which she was raised (396).

For sure, Klein acknowledges, the sun, wind, and waves cannot be "possessed" by us like coal or oil. But renewables are the only way to avoid the catastrophe of climate change; by choosing them "we find ourselves back where we started, in dialogue with nature." "Back" has here a very precise meaning: before James Watt's steam engine, which supposedly liberated us – or rather liberated factory owners – from the need to adapt to the rhythms of natural ecosystems (394). This choice is not a simple "technical" matter, but a cultural and political struggle in which the young Cheyenne followers and students of Red Cloud are on the front line, as "solar warriors" (397).

Klein sometimes describes this struggle as one between an "extractivist mindset" and a "regenerative" one. For us, it might be more appropriate to speak of a confrontation between two opposed ways of life, as well as worldviews. Often the most articulate exponents of this "regenerative" culture are women, like the brilliant Mississauga Anishinaabe writer Leanne Simpson, who has devoted much of her life to "collecting, translating and artistically interpreting her people's oral histories and stories." According to her, the Anishinaabe have "a way of living designed to generate life, not just human life but the life of all living things." Klein comments: "This is a concept of balance, or harmony, common to many Indigenous cultures" in confrontation with "the bulldozers of colonialism and corporate globalization." The Anishinaabe words for it are often translated to mean "the good life," but Simpson prefers the translation "continuous rebirth," which she heard from fellow Anishinaabe writer and activist Winona LaDuke (442–443).

Curiously, despite her feminist commitment Klein does not discuss the growing role of women in these social and ecological indigenous struggles – a role that does not result from some "essential" affinity with Mother Earth but from the simple fact that women, by virtue of their traditional occupations with land, food, and water, are most directly affected by the environmental disasters produced by capitalist "modernization."

Protest against the mercantile/monetary quantification of the world in the name of past qualitative values, a key component of the romantic anti-capitalist worldview, can be found in many pages of *This Changes Everything*, always linked to the issue of the natural environment, i.e. to nature-protective traditions that resist the "icy waters of egotistical calculation," in Marx's famous phrase from *The Communist Manifesto*. For example, mentioning the refusal of "compensations" for the Northern Gateway Pipeline by Jess Housty, a female spokesperson for the Heiltsuk indigenous community, Klein comments:

> When what is being fought for is an identity, a culture, a beloved place that people are determined to pass on to their grandchildren, and that their ancestors may have paid for with great sacrifice, there is nothing companies can offer as a bargaining chip. No safety pledge will assuage; no bribe will be big enough (342).

A similar argument is voiced by Lucas King, a young Cheyenne leader, explaining to a panel of unsympathetic officials the refusal, by his community, of the "benefits" of a large coal mine: "This is Cheyenne country. It has been for a long time, longer than any dollar has ever lasted. I don't expect you guys to understand us. You don't." (397).

In other words, there is an essential – one might say ontological – incompatibility, or incommensurability, between these ancestral values and the purely quantitative measures of the capitalist market. As Klein strikingly puts it: "no amount of money" can extinguish the "ferocious love" of indigenous communities for their lands and waters (342). These communities are usually poor, but they are determined to "defend a richness that our economy has not figured out how to count" (344) – in fact it cannot "account" for it because it does not belong to the measurable world of commodities.

Water, the source of life, is a particularly sensitive issue. A Romanian villager resisting fracking asserts: "We are not even that poor. Maybe we don't have money, but we have clean water and we are healthy and we just want to be left alone." According to Klein, environmental battles are often grounded on a stark choice: "Water vs gas. Water vs oil. Water vs coal." The anti-fossil fuel movements are above all "pro-water movements" (344). Charlene Alden, the tireless director of the Cheyenne environmental protection office, explains why she refuses to "sacrifice the health of land for coal dollars": "in Cheyenne, the word for water is the same as the word for life ... We know that if we start messing around too much with coal, it destroys life." By choosing solar power, Charlene Alden, Red Cloud, and other activists, have been able to prove to the new generation of Cheyennes that "there is another path out of poverty and hopelessness – one that does not involve handing over the land for which their ancestors paid so dearly" (391).

One of the most attractive features of *This Changes Everything* is the way Klein gives indigenous spokespersons and activists – often women, an emerging force everywhere – the opportunity to explain – with their own words, in the language of their culture and tradition – their way of life and their battle against the bulldozers of "extractivism," driven by the fossil fuel oligarchy. But fortunately, indigenous communities are not alone in this struggle. They are often joined by peasant and farmer movements – Via Campesina, for example, a large international network fighting agribusiness that might have deserved even more attention in the book – as well as trade-unionists, students, ecologists, various radical activists, and of course the local non-indigenous population. Large coalitions are thus built around indigenous rights, such as "Save the Fraser River," which unites more than 130 First Nations and many non-indigenous endorsers against the disastrous Canadian Northern Gateway Pipeline (345).

According to Klein, a surprising and very important development has taken place in recent years:

> many non-Natives are also beginning to see that the ways of life that Indigenous groups are protecting have a great deal to teach about how to relate to the land in ways that are not purely extractive. This represents a true sea change over a very short period of time (371).

Klein is not arguing, like certain "primitivist" ecologists, that everybody should return to the indigenous way of life, but rather that there are important lessons to be learned from them, lessons that are crucial for the building of alternatives to the suicidal course of the modern way of life.

Thanks to the common struggle in Blockadia, and the leadership roles taken by indigenous people, "Indigenous-inspired ideas have spread" in a way "that has not occurred for centuries." Not only in Canada and the United States, but all around the world, "many people are remembering their own cultures' stewardship traditions, however deeply buried, and recognizing humanity's role as one of life promotion." Confronted with the growing threat from the aggressive search for "non-conventional" fossil fuel resources – an "extreme energy frenzy" – "these old ideas are reasserting themselves – cross-pollinating, hybridizing, and finding applications in new contexts" (443–445).

It might be asked whether Klein isn't being overly optimistic, and underestimating the power of the system, the insidious force of the "Almighty Dollar" (John Muir). The most important thing, in our view, is that she has perceived, with remarkable insight, the critical, subversive, and utopian power of the "ancient", the "old," and the "traditional," both for resisting the perverse logic of the present social and economic order, and for imagining a radically new one.

Time to put on the brakes

Klein may be optimistic, but she is keenly aware of the danger that climate change represents for humanity. In *This Changes Everything*, she calls it "one of history's most damned crimes" (464). Our economic system, she says, is "at war with many forms of life on earth, including human life" (21). She is also aware of the formidable coalition of economic and political forces interested in maintaining, at whatever price, this economic order. And, last but not least, she understands that the time for gradual reforms, "polite incremental change," is over. Only a systemic approach, a process of radical change, can save us:

> the things we must do to avoid catastrophic warming are no longer just in conflict with the particular strain of deregulated capitalism that triumphed in the 1980s. They are now in conflict with the fundamental imperative at the heart of our economic model: grow or die (21).

In this passage, then, she indicates clearly that her thinking at the deepest level has gone beyond Keynesianism.

In other words, Klein believes, using words strikingly similar to those of Walter Benjamin, that it is time "to put the brakes on the forces of destruction" (451). Where, though, is hope to be found? What are the forces able to undertake such a profound transformation, one which Benjamin defined as revolutionary? Klein quotes a speech, not by a radical ecologist at a leftist meeting, but by Brad Werner, an American scientist, at the Fall Meeting of the American Geophysics Union, in San Francisco, in December 2012. As Klein reports his words, the only hope he sees for saving the "earth-human systems" is "'resistance' – movements of 'people or groups of people' who 'adopt a certain set of dynamics that does not fit within the capitalist culture'." According to Werner, this sort of resistance includes "environmental direct action, resistance taken from outside the dominant culture, as in protests, blockades and sabotage, by Indigenous peoples, workers, anarchists and other activist groups." Put another way, Klein adds, "only mass social movements can save us now. Because we know where the current system, left unchecked, is headed" (450). The imperative is to move beyond "fleeting pockets of liberated space" towards the building of a new world: "the stakes are simply too high and time too short to settle for anything less" (465).

Is the radical systemic change that is needed some form of socialism, or eco-socialism? *This Changes Everything* avoids the term, but the book advocates for public ownership and long-term planning (94, 99), two key components of any socialist project. The emphasis in Klein's vision of the future is however on the ethical foundations of the alternative way of life: "unafraid of the language of morality," she does not hesitate to "give the pragmatic, cost-benefit arguments a rest and speak of right and wrong, of love and indignation." Her historical model is the uncompromising fight for the abolition of slavery. But she believes that the same understanding about the need to assert the value of life which inspired the abolitionists:

> is at the heart of all major progressive victories, from universal suffrage to universal health care. Though these movements all contained economic arguments as part of building their case for justice, they did not win by putting a monetary value on granting equal rights and freedoms. They won by asserting that those rights and freedoms were too valuable to be measured and were inherent to each of us (463–464).

The same argument is valid for the issue of climate change: "cost-effective" calculations are "morally monstrous," since they imply that "there is an acceptable price for allowing entire countries to disappear" (464).

Right and wrong, love and indignation, life and freedom versus monetary calculations: are we not here at the heart of the romantic worldview? In *This Changes Everything* Klein seems to suggest that such a perspective, in new and unexpected forms, is not irrelevant for the challenges of the twenty-first century.

No is Not Enough: utopia and memory

Although *This Changes Everything* represents her fullest and most elaborated statement of a romantic ecology, insisting on the urgency and centrality of the question of climate change, Klein has by no means abandoned this perspective in what she has written since. Clearly the shift of focus is a lasting one. This was evident already in a talk she gave in 2016 – The Edward W. Said London Lecture – which was first published in the *London Review of Books* in the same year, then republished in a collective volume in 2017.[13] Its title, "Let Them Drown: The Violence of Othering in a Warming World," suggests the intimate connection she feels exists between the ecological crisis and the endangerment of the way of life of oppressed, devalued, "othered" peoples. The fact that she makes climate change and ecological damage the centerpiece of a talk in honor of Edward Said is at first sight paradoxical, as she points out at the outset, since Said was not, and had little time for, what he sometimes called "tree-huggers." But Klein goes on to demonstrate the relevance both of ecological issues for Said's main concerns, and, most importantly, of Said's work for the ecological movement.

For Said was the great theorist and historian of "othering," and with climate change and the ravages of modern industrialism more generally, areas of colonization, external or internal, inhabited by "backward" and more "primitive" peoples, come to be defined as "sacrifice zones" by the wealthy and powerful, in the name of progress and profit. But, Klein emphasizes, this intense, combined exploitation of nature and of subordinated peoples is specific to the modern, capitalist era, and she makes an important point about the recently coined term "Anthropocene":

> We often hear climate change blamed on "human nature," on the inherent greed and short-sightedness of our species. Or we are told we have altered the earth so much and on such a planetary scale that we are now living in the Anthropocene – the age of humans. These ways of explaining our current circumstances have a very specific, if unspoken meaning: that humans are of a single type, that human nature can be essentialized to the traits that created this crisis. In this way, the systems that certain humans created, and other humans powerfully resisted, are completely let off the hook ... Diagnoses like this erase the very existence of human systems that organized life differently ... These systems existed and still exist, but they are erased every time we say that ... we are living in the "age of man."[14]

The systems that organized life differently were of course pre-modern, pre-capitalist ones, or communities subsisting in modernity that to some extent preserve such patterns of living.

In her latest book to date, *No is Not Enough*, Klein further develops the idea, based on her reporting on and participation in current ecological acts of resistance, that indigenous communities still in touch with earlier modes of social and natural relations are key to the struggle against the catastrophes wrought by contemporary capitalism. *No is Not Enough* was written more rapidly than her previous books, as an active, immediate contribution to countering the threat represented by Trump's election to the presidency in the United States and the rise of demagogic nationalism elsewhere; much of it involves analysis of the present socio-political situation so created. But in its final section it returns to the themes of romantic ecology first fully elaborated in *This Changes Everything*.

In her remarkable discussion of Trump and what his electoral victory signifies, Klein refuses to reduce the issue to one of personal psychopathology and sees Trump as the culminating symptom of a whole civilization. Drawing on her analysis of branding in *No Logo*, she perceives him to incarnate the perfect fusion of self and brand; as such he is the "Hollow Man," not an isolated monstrosity but on the contrary a reflection of the "profound emptiness at the heart of the very culture that spawned [him]."[15] The hollowness of modern culture comes from its elimination or stunting of "community, connection, a sense of mission larger than our immediate atomized desires;" brands, which cater to these desires, at the same time offer people the illusion of "being part of something larger than ourselves" but are unable to sustain it because they cannot authentically fulfill that yearning.[16]

The larger "neoliberal" economic trends that have dominated at least since the 1980s and that Trump purveys in spite of his populism, have brought both social injustices and ecological devastations to a crisis pitch, and in the face of impending catastrophe Klein sees, as in *This Changes Everything*, the need for a radical break. In *No is Not Enough* the metaphor changes slightly, but the idea is the same: her "reporting from disaster zones" has given her "a preview of where the road we are all on is headed unless we somehow grab the wheel and swerve."[17]

In the final section of her book Klein addresses the question of how this might be done. Its epigraph is a quotation from the Uruguayan writer Eduardo Galeano, in which he rhetorically asks what the good of utopia is, when it continually recedes on the horizon as one walks towards it. The answer – "it's good for walking" – underlines the crucial role of utopian thinking in motivating rebellious movements of social change.[18] In one particularly significant passage in this last section, Klein expands on that proposition. In answer to the question of why the period of the early twenty-first century produced little resistance to the calamitous status quo, she suggests that what was lacking overall was the utopian impetus: "generations who had grown up under neoliberalism struggled to picture something, anything, other than what they had always known." To explain why that was so, Klein makes an interesting hypothesis, linking the utopian dynamic with memory. As an illustration of this connection, she cites the workers' movements of the nineteenth century:

> When workers rose up against the depravities of the industrial age, many had living memories of a different kind of economy. Others were actively fighting to protect an existing way of life, whether it was the family farm that was being lost to predatory creditors or small-scale artisanal businesses being wiped out by industrial capitalism. Having known something different, they were capable of imagining – and fighting for – a radically better future.[19]

Although Klein makes no overt reference to him, this passage has clear overtones of E.P. Thompson's work on the "moral economy" of the English working class.[20]

In the present situation in North America and the West more generally, Klein continues, there are nonetheless some exceptions to the loss of memory of an alternative reality that leads to the inability to envision utopia: "There are specific cultures and communities – most notably Indigenous communities – that have vigilantly kept alive memories and models of other ways to live, not based on ownership of the land and endless extraction of profit."[21] Klein has observed the joining of memory and utopia in indigenous communities first hand, and she devotes a full chapter to this experience: "Lessons from Standing Rock: Daring to Dream." In December 2016,

Klein joined the camp of "water protectors" near the Standing Rock Sioux Reservation that were attempting to block the passage of the Dakota Access pipeline through that territory – a passage that would at the same time endanger the natural habitat and desecrate lands held sacred by the Sioux. In talking with indigenous leaders there, she was deeply impressed by the importance of "traditions and ceremonies that had been kept alive despite hundreds of years of genocidal attacks ..."[22] She was equally stirred by the closeness – even near-identity – of the Amerindians with their natural environment; in that relation, as she describes it, the notion of protection is essential, and it is a fully reciprocal one – nature protects indigenous people, and they protect it.[23]

Near the end of *No is Not Enough*, in attempting to define the positive – and utopian – outlook she considers necessary to inspire the movement of opposition to neoliberal capitalism, Klein calls for "a shift in values, indeed in morality" that would replace the amorality of exploitation.[24] Although she discusses the needed projection of a desirable future in more concrete terms as well, this broad ethical concept seems to be at the core of her vision. As we have seen, in developing this moral perspective Klein has been strongly affected and influenced by the value systems – in particular as they relate to the natural world and are epitomized in such notions as "balance," "stewardship," and "protection" – of the indigenous peoples she has encountered. In this light, Klein appears clearly as a contemporary – and illustrious – descendant of the tradition of romantic anti-capitalist ecology we have explored in this book.

Notes

1 See Klein, *The Shock Doctrine: The Rise of Disaster Capitalism*, 20.
2 See Klein, *No is Not Enough: Defeating the New Shock Politics*, Part IV: "How Things Could Get Better."
3 Klein, *No Logo*, 68.
4 Klein, *No Logo*, 154–156.
5 The introduction to *The Shock Doctrine* is entitled "Blank is Beautiful: Three Decades of Erasing and Remaking the World."
6 Klein, *Fences and Windows: Dispatches from the Front Lines of the Globalization Debate*, 30, 33.
7 Klein, *No Logo*, Chap. 13, esp. 313, 323.
8 Klein, *Fences and Windows*, 67.
9 Klein, *No Logo*, 425.
10 Klein, *The Shock Doctrine*, 443.
11 Klein, *Fences and Windows*, 222, 223. Klein also contributed a text to the collective volume by Subcomandante Marcos, *Ya Basta! Ten Years of the Zapatista Uprising* (2004).
12 Klein, *This Changes Everything: Capitalism vs. The Climate*, 157. All further page references to this work will be indicated parenthetically in the text.
13 *Will the Flower Slip Through the Asphalt: Writers Respond to Capitalist Climate Change*. Edited by Vijay Prashad.
14 Klein, "Let Them Drown: The Violence of Othering in a Warming World," in *Will the Flower Slip Through the Asphalt*, 39–40.
15 Klein, *No is Not Enough*, 59.

16 Klein, *No is Not Enough*, 60.
17 Klein, *No is Not Enough*, 131.
18 Klein, *No is Not Enough*, 187.
19 Klein, *No is Not Enough*, 219.
20 See especially Thompson, *The Making of the English Working Class* and *Customs in Common*.
21 Klein, *No is Not Enough*, 220.
22 Klein, *No is Not Enough*, 225.
23 Klein, *No is Not Enough*, 226–227.
24 Klein, *No is Not Enough*, 240–241.

Works Cited

Klein, Naomi. *Fences and Windows: Dispatches from the Front Lines of the Globalization Debate*. London: Harper Perennial, 2005 (1st edition: 2002).

Klein, Naomi. *No is Not Enough: Defeating the New Shock Politics*. London: Allen Lane, 2017.

Klein, Naomi. *No Logo*. London: Fourth Estate, 2010 (1st editions: 1999 and 2000).

Klein, Naomi. *The Shock Doctrine: The Rise of Disaster Capitalism*. London: Penguin, 2008 (1st edition: 2007).

Klein, Naomi. *This Changes Everything: Capitalism vs. The Climate*. New York: Simon & Schuster, 2014.

Thompson, Edward Palmer. *Customs in Common*. New York: Norton, 1993.

Thompson, Edward Palmer. *The Making of the English Working Class*. London: Victor Gollancz, 1963.

Will the Flower Slip Through the Asphalt: Writers Respond to Capitalist Climate Change. Edited by Vijay Prashad. New Delhi: LeftWord Books, 2017.

Ya Basta! Ten Years of the Zapatista Uprising. Edited by Ziga Vodovnik. Oakland, CA: AK Press, 2004.

Conclusion

In this book we have studied a series of figures who exemplify what we have called romantic ecocriticism, that is, an ecological awareness, concern, and criticism that comes specifically out of a romantic anti-capitalist perspective. We are aware that on the one hand some manifestations of ecology or environmentalism are not romantic in this sense, their viewpoint not always incompatible with acceptance of capitalism, while on the other hand some romantic sensibilities are relatively indifferent to the natural world and ecological questions. But what has interested us here is to explore the meeting of romantic anti-capitalism and ecological consciousness, in what we believe to be an "elective affinity" between the two.

The social origins of the figures we have discussed are by no means homogeneous, showing that the general perspective they have in common is not determined in any mechanistic way by social class. William Morris, one of the most radical voices, paradoxically came from the wealthy English upper middle class, and most of the other figures have middle-class or bourgeois roots of some kind. It is noticeable, though, that usually their position is marginal or relatively subordinate in relation to or within the ruling elites. William Bartram's family were botanists engaged in some commercial activity rather than full-fledged merchants; Thomas Cole's father was a failed businessman; Walter Benjamin came from a German Jewish family; and Naomi Klein's parents were middle-class bohemian intellectuals, Jewish as well. Raymond Williams stands out, though, in coming from a Welsh working-class family, thus doubly distanced from the dominant classes of English society.

Our study has covered a very broad time span, one that runs from the inception of capitalist modernity to its latest, present-day manifestations, and our subjects thus confront it in different stages and locations: Bartram in the early development of a fully market-dominated society in British America; Cole in the beginnings of industrialization in "Jacksonian America;" Morris in late nineteenth-century England, at the height of industrial capitalist expansion; Benjamin during the crises of capitalism in early twentieth-century Europe; Williams in post-World War II Britain; and Klein, finally, who comes of age in the most recent era of "branding" and "neoliberalism" in North America. In these different stages and situations, the responses of the figures treated here show a certain pattern of evolution. Whereas Bartram and Cole were largely powerless in their relative isolation, voices crying in the wilderness and to a certain extent coopted in the end, starting with Morris forms of collective activism became possible in the struggle to oppose the modern forces of ecological depredation. Williams, like Morris, joined in group actions, although his principal form of intervention remained writing. Benjamin is the exception to this trend, since his form of action was purely intellectual (which of course does not make it any the less effective ultimately), while Klein represents the culmination of the tendency, with her extensive participation in ecological organizations and demonstrations.

In spite of these differences and variations, as well as those resulting from the wide diversity of their chosen fields of activity and creation, our figures are united in sharing the romantic worldview in its ecocritical variant. While Morris is doubtless the one who comes closest to incarnating the Weberian ideal type of this perspective, all of them are exemplary cases that express it in radical form. In one way or another, they often stand out in the context of their time and place. In addition to Bartram and Cole, whose exceptional character we have emphasized, we have noted that Benjamin can be seen as the first Marxist thinker to break with the ideology of "progress," and Williams as the first British socialist to adopt an ecocritical position. Despite this exemplarity, powerful external pressures could in some instances lead to a degree of ambiguity or contradiction in their viewpoints, without however undermining their fundamental vision. Bartram and Cole, for instance, sometimes articulated their critical perceptions more clearly and uncompromisingly in their unpublished than in their published writings; and in the case of Benjamin, his *rapprochement* with the Soviet productivist model was limited to a short period, 1933–1935, even then not affecting all of his work.

While we have thus highlighted the unique status and contribution of each of them, we have also shown that they are part of a continuous, larger tradition of romantic ecological concern and protest, and notably have linked them to the literary figures that have been the principal subject of study of "ecocriticism," as discussed in our introduction. We have seen how our subjects had significant interrelations and interactions with the long English literary tradition studied by Williams, running from Blake, Wordsworth, Coleridge, and John Clare, through Carlyle, Cobbett, and Ruskin, into the twentieth-century writers who came out of that cultural matrix. We have also alluded to links – more or less close – with the related American strand of nature writing, of which Thoreau, Emerson, and John Muir are the most illustrious exponents; with early German romanticism and with the pioneering German ecological scientist-traveler Alexander von Humboldt; finally, and crucially, we have (in the introduction) underlined the seminal importance of Rousseau at the beginnings of romantic ecocriticism.

The impact our subjects within this broad tradition has often involved carrying forward and developing in new ways their trenchant insights and the integrity of their romantic vision. As we have pointed out, though, in some cases interpretation of their work by others whose perspective is less sharply oppositional has edulcorated it, blunting or defusing the critical force of the original, as for example with Cole and proponents of the National Parks movement, or Morris and the Garden Cities, or yet again some post-modern readings of Benjamin. Yet the radical nature of their vision has been evident, and inspirational, to many.

In the discussions of these figures, lines of interaction have also emerged between their romantic anti-capitalism and other ideological constituents. Some of these were religious, as with Bartram's relation to the "primitive" strand of Quakerism or Benjamin's to messianic Judaism. Others were political, as with Cole's ties to "aristocratic" Whiggism, Morris's and Benjamin's, in different ways, to anarchism, and those of the latter two, along with Williams, to Marxism. In the case of Benjamin the perspective of a cultural movement – Surrealism – has also informed his work. These crossings with multifarious other ideological elements further illustrate the diversity that exists within romantic ecocriticism.

The common defining characteristics of all of the figures we have analyzed, as we have clearly seen throughout, are a passionate critique of modern civilization, in the name of values rooted in a pre-modern era, with a special emphasis on the question of its relation to the natural world. While they often express similar perceptions in each of these three aspects of the romantic ecocritical vision, they also sometimes focus on different issues and phenomena, or articulate their ideas through different concepts. In critiquing modernity, for example, Cole aims at the reigning ideology of "utilitarianism," Benjamin at the mechanical nature of modern society, Williams at what he calls "productivism" and "progressivism," Klein at the hollowing out of contemporary society through branding. Central to all of their discourses is a subversive questioning of the very definition of wealth in that society, and all in their own way propose a redefinition of what real wealth is, showing how it is negated and destroyed in modernity.

The periods of the past in which they find the locus of true values in opposition to modernity varies, although we can notice a marked propensity for one broad epoch in particular. Although attraction for the Middle Ages is quite common among romantics otherwise, Morris is alone among our subjects in expressing this predilection. While one of Cole's preferred social states of the past is a broadly defined one of pastoralism and small farming, he is at the same time drawn to "primitive" societies in wilderness settings, and this is the past most valued by our subjects as a whole, from Bartram's Amerindians through Benjamin's "*Urzeit*" of primitive matriarchy as delineated by Bachofen, to Klein's contemporary "indigenous" peoples. We will return later to the privileged status of the latter with regard to romantic ecologism.

All of our subjects project a strong emotive identification with the world of "nature," however experienced and perceived, and are painfully aware that it is damaged and endangered in modernity, as it was not in the past periods they value. Their specific conceptions of nature vary, but often share several common themes. They often see it as fully egalitarian, in which no hierarchy exists between the multitude of life forms, and also as entirely integrated, interconnected, unified in an all-encompassing network. In this way it stands in stark contrast with the forms of domination and fragmentation in modern society. Beyond these characteristics, though, and more specifically romantic in nature, are the magic, enchantment, and spirituality that nature is seen to possess. Finally, in the Western cultural universe, the metaphor of the mythical Garden of Eden is evoked again and again.

Their appraisal of the ills generated by capitalist modernity often lead those we have studied to a sense of calamitous crisis that expresses itself in apocalyptic, dystopian imaginings, and prophetic warnings. Such premonitions of impending doom, unless significant change occurs, are an important component, especially, of Cole's work – his writing as well as his great series of paintings, *The Course of Empire*. But they are present also in Benjamin's anguished metaphors of cutting the fuse before it reaches the explosive, and putting on the brakes before the crash occurs. Klein, who is acutely conscious of the ultimate peril that climate change represents, though probably not aware that Benjamin had used the image before her writes of the absolute, urgent need to apply the brakes.

The romantic ecocritics, then, are often deeply pessimistic in their analyses and pronouncements. Yet in some cases theirs is – as we termed it with regard to Benjamin – a "revolutionary pessimism," one that refuses despair, dares to imagine utopia and sometimes to work towards it. The utopian vision is, in fact, a potent presence beside the dystopian. It is most fully manifest, of course, in Morris's *News from Nowhere*, but it can be found elsewhere in different guises: in Bartram's and Cole's evocations of communities living as *part* of nature; in Williams's memories of "knowable communities," and yearning for the transcendence of the city/country opposition; in Klein's recognition that utopian thinking is essential today, and her insight that it is linked to *memory*. As for the paradox of the coexistence of pessimism and optimism in our subjects, perhaps the last word should be Benjamin's: every second is the narrow gate through which salvation can come.

We believe that the intellectual, emotional, and cultural relation between the romantic protest against capitalist modernity and care for the natural environment – as documented in this book – is not a thing of the past but quite relevant for the present and doubtless the future as well. Our discussion in the last chapter of Klein's writings has highlighted one important instance of its continued presence today. The connection can also be found in some recent and contemporary literary authors – two notable examples are the Native American poet and fiction writer Leslie Marmon Silko and the science fiction fantasy novelist Ursula Le Guin – and in cinema as well. A striking example of the latter is the film *Avatar* (2009) by James Cameron, a futuristic fantasy that portrays the violent confrontation between human colonizers – bent on extracting a precious mineral that will allow them to resolve the energy crisis caused by the severe depletion of the Earth's resources – and the Na'vi, a humanoid species indigenous to the planet Pandora, living in harmony and symbiosis with nature. The conflict is seen through the eyes of one of the colonizers who finally decides – perhaps at least partially predisposed by his physical disability – to take sides with the indigenous people and become one of them.

The romantic perspective is also clearly present in some of the more radical currents of the ecological movement: "deep ecology," eco-feminism, ecological primitivism, and eco-socialism. And even though Thomas Cole may have no real equivalent in the twenty-first century, there are Romantic artists with "naturalist" sensibilities also in our time.[1] To discuss these manifold manifestations would of course require a new book, so we will focus here, in conclusion, on a broad cultural phenomenon that we see as crucial to present and future developments in the ecological arena, the one that forms the background for the romantic vision of Naomi Klein: *indigenism*. This term has many and contradictory meanings, but the most relevant from our perspective refers to social and political movements aiming at the emancipation – or self-emancipation – of the indigenous populations of the Americas. It first appeared with the Mexican Revolution of 1911–1917, and was given a Marxist interpretation in the writings of the Peruvian socialist José Carlos Mariátegui in the late 1920s. It became widespread in the twenty-first century, and has even been adopted as official policy by some leftist governments, notably that of Evo Morales in Bolivia.

As already discussed in relation to Naomi Klein, indigenous communities worldwide, but especially in the Americas, have recently fought in defence of their land, water, and forests, against pipelines, gold mines, oil companies, and industrial agro-business. They have in fact become central in the struggle for the environment. This is true not only through their local actions, but also in that they propose an alternative way of life to that of neo-liberal globalized capitalism. The political culture of this "indigenism," as expressed in documents of the movements and of their supporters in the broader society – songs, prayers, literary works, films, and theoretical essays – has a powerful romantic and ecological dimension, since in this perspective criticism of the modern devastation of "Mother Earth" is often inspired by pre-capitalist communitarian traditions and spiritual values, and a relation to nature as a sacred realm.

The dynamics of capital require the transformation of all commonly held goods into commodities, which sooner or later leads to destruction of the environment. The petroleum zones of Latin America, abandoned by the multinationals after years of exploitation, are left poisoned and gutted, with a dismal legacy of illness among the inhabitants. The populations that live in the most direct contact with these environments are the first victims of this ecocide and understandably attempt to oppose the destructive expansion of capital on their lands as a matter of survival. Beyond these immediate, concrete motivations, though, there comes into play a deeper antagonism between the way of life of these communities and the "spirit of capitalism" as Max Weber defined it. There exists a kind of "negative affinity" between indigenous ethics and the spirit of capitalism – the converse of the elective affinity between the Protestant ethic and capitalism.

This conflict has a long history. It was already (admirably) described in one of the Mexican novels of the anarchist writer B. Traven, *The White Rose* (1929), which chronicled the seizure by a large North American oil company of the lands of an indigenous community after the murder of its leader. But the conflict has intensified during the last few decades, due on the one hand to the increased intensity and extensiveness of exploitation of the environment, and on the other to the rise of the Global Justice movement combatting it alongside indigenous movements. Among the best known examples, some of which appear in Klein's *This Changes Everything*, are: the struggle of Canadian First Nations against the exploitation of tar sand oil in Alberta; the Sioux tribal resistance to the Dakota Access Pipeline; the indigenous Brazilian movement for an "oil-free Amazon forest;" the opposition by Peruvian communities in the Cajamarca area to poisonous gold mining; and the Ecuadorian indigenous movement's combat, supported by ecologists, against the concessions to oil multinationals in the Yasuni Park. Similar movements, some also dealt with in *This Changes Everything*, have taken place elsewhere in the world.

In April 2010, a People's Conference on Climate Change and the Rights of Mother Earth was convened by Evo Morales in the Bolivian city of Cochabamba. More than 20,000 delegates from all over the world took part, but the majority were from the Andean countries of Latin America, with a very substantial indigenous representation. The resolution adopted by the conference, which has had considerable international repercussions, is an impressive document of the "romantic" ecological and anti-capitalist idea of indigenism. The following are excerpts from this document:

> The capitalist system has imposed on us a logic of competition, progress and limitless growth. This regime of production and consumption seeks profit without limits, separating human beings from nature and imposing a logic of domination upon nature, transforming everything into commodities: water, earth, the human genome, ancestral cultures, biodiversity, justice, ethics, the rights of peoples, and life itself.
>
> Under capitalism, Mother Earth is converted into a source of raw materials, and human beings into consumers and a means of production, into people that are seen as valuable only for what they own, and not for what they are.
>
> Capitalism requires a powerful military industry for its processes of accumulation and imposition of control over territories and natural resources, suppressing the resistance of the peoples. It is an imperialist system of colonization of the planet.
>
> Humanity confronts a great dilemma: to continue on the path of capitalism, depredation, and death, or to choose the path of harmony with nature and respect for life.
>
> It is imperative that we forge a new system that restores harmony with nature and among human beings. And in order for there to be balance with nature, there must first be equity among human beings. We

> propose to the peoples of the world the recovery, revalorization, and strengthening of the knowledge, wisdom, and ancestral practices of Indigenous Peoples, which are affirmed in the thought and practices of "Living Well" (*Suma Kawsay*), recognizing Mother Earth as a living being with which we have an indivisible, interdependent, complementary and spiritual relationship.[2]

A separate resolution was adopted by the indigenous delegations at the same conference. Here is a significant passage from that text:

> We Indigenous Peoples are sons and daughters of Mother Earth, or "Pachamama" in Quechua. Mother Earth is a living being in the universe that concentrates energy and life, while giving shelter and life to all without asking anything in return, she is the past, present and future; this is our relationship with Mother Earth. We have lived in coexistence with her for thousands of years, with our wisdom and cosmic spirituality linked to nature. However, the economic models promoted and forced by industrialized countries that promote exploitation and wealth accumulation have radically transformed our relationship with Mother Earth. We must assert that climate change is one of the consequences of this irrational logic of life that we must change ... The aggression towards Mother Earth and the repeated assaults and violations against our soils, air, forests, rivers, lakes, biodiversity, and the cosmos are assaults against us.[3]

Of course the contemporary "indigenism" expressed here in the terms of an activist political movement is quite different from the individual nineteenth- and twentieth-century versions of the ecological romantic worldview discussed in this book, yet an underlying communality is clearly visible. As illustrated here, romanticism as we conceive it – generally as well as in its specifically ecological manifestation – can be expressed not only in individual authors or artists but also in social movements. In the case of ecological romanticism, the work of Naomi Klein can be seen as forming a bridge between the two. Although the future cannot be predicted with any certainty, in view of the fundamental and universal urgency of the climate change crisis that now confronts us it is to be hoped that romantic anticapitalist ecocriticism will continue to make its forceful voice heard in the anti-systemic rebellions of the future.

Notes

1 One example, which brings together art, ecology, romanticism, and critical theory, is the illustrated album *Truth and Dare, a Comic Book Curriculum for the End and the Beginning of the World*, published in 2014 by ecosocialists Fred Ho and Quincy Saul.

2 See World People's Conference on Climate Change and the Rights of Mother Earth (website). Accessed April 9, 2019. https://pwccc.wordpress.com/2010/04/24/peoples-agreement.
3 See World People's Conference on Climate Change and the Rights of Mother Earth (website). Accessed April 9, 2019. https://pwccc.wordpress.com/2010/04/24/peoples-agreement. Working Groups: 07. Indigenous Peoples.

Index